A PREHISTORY OF THE NORTH

A PREHISTORY OF THE NORTH

*Human Settlement
of the Higher Latitudes*

John F. Hoffecker

RUTGERS UNIVERSITY PRESS
New Brunswick, New Jersey, and London

Library of Congress Cataloging-in-Publication Data

Hoffecker, John F.
   A prehistory of the north : human settlement of the higher latitudes / John F. Hoffecker.
      p. cm.
   Includes bibliographical references and index.
   ISBN 0-8135-3468-2 (hardcover : alk. paper) —
   ISBN 0-8135-3469-0 (pbk. : alk. paper)
      1. Neanderthals—Arctic regions.   2. Prehistoric peoples—Arctic regions.
   3. Human beings—Arctic regions—Migrations.   I. Title.

GN285.H65 2005
930′.091—dc22

                                                                          2004000306

A British Cataloging-in-Publication record for this book is available from the British Library.

Copyright © 2005 by John F. Hoffecker
All rights reserved
No part of this book may be reproduced or utilized in any form or by any means, electronic or mechanical, or by any information storage and retrieval system, without written permission from the publisher. Please contact Rutgers University Press, 100 Joyce Kilmer Avenue, Piscataway, NJ 08854–8099. The only exception to this prohibition is "fair use" as defined by U.S. copyright law.

Manufactured in the United States of America

In memoriam
William Roger Powers
(1942–2003)

Contents

Foreword *by Brian Fagan*   ix
Preface   xiii

CHAPTER 1: Vikings in the Arctic   1
CHAPTER 2: Out of Africa   10
CHAPTER 3: The First Europeans   28
CHAPTER 4: Cold Weather People   47
CHAPTER 5: Modern Humans in the North   70
CHAPTER 6: Into the Arctic   96
CHAPTER 7: Peoples of the Circumpolar Zone   120

Notes   143
Bibliography   183
Figure Credits   211
Index   215

Foreword

The stereotypes of the Cro-Magnons have long been with us—reindeer hunters, cave dwellers, and consummate artists—people who adapted brilliantly to the harsh environmental realities of the late Ice Age world 18,000 years ago. Cro-Magnons were indeed expert cold-weather hunter-gatherers, but their success is only part of a much larger and, until recently, little-known story. *Prehistory of the North* explores human adaptations to the colder latitudes of the world on a much broader canvas, and in so doing fills a huge gap in our knowledge of human history.

John Hoffecker paints a broad-brushed picture, based in part on his own researches, also on an encyclopedic knowledge of a huge specialist literature culled from obscure journals and monographs in many languages. He ranges from human origins in tropical Africa to the spread of archaic and modern humanity into middle latitudes to Norse voyages to Greenland and beyond. He explores the first colonization of Europe and Eurasia—the limited success of archaic humans like the Neanderthals, then the conquest of the great Eurasian steppe-tundra by *Homo sapiens sapiens*. This is a book about adapting to environments with extremes of seasonal temperatures and low productivity, about arctic deserts that suck in and expel their human inhabitants. The issues are myriad—physical and behavioral adaptations to cold environments, diets high in fat and protein from game animals, and the need for adequate shelter in subzero winters to mention only a few. By deploying evidence from numerous scientific disciplines, Hoffecker describes the compelling reasons why archaic humans never settled year-round in the world's most demanding environments.

These colder regions, all in the Northern Hemisphere, were beyond the Neanderthals and their contemporaries, who settled in such areas as southwestern Siberia but never established permanent hunting territories on the open plains to the north. These inhospitable latitudes were the provinces of modern humans, who spread into late Ice Age Europe and rapidly across the steppe-tundra after 45,000 years ago. With fully developed cognitive abilities, sophisticated linguistic skills, and an ability to plan ahead, to conceptualize their world, *Homo sapiens sapiens* soon mastered the north. Highly mobile, armed with a very sophisticated technology that included the eyed needle and the layered, tailored clothing made possible by it, our Ice Age ancestors had colonized much of Eurasia by 25,000 years ago, before the last cold snap of the Weichsel glaciation that climaxed 18,000 years ago. Hoffecker marshals what little we know about the earliest inhabitants of the steppe-tundra with its nine-month winters from an archaeological signature that is always exiguous, always tantalizing. He shows how these pioneers, and their successors, were highly sensitive to extreme cold, moving southward into more sheltered terrain during glacial maxima, moving northward again with their prey as conditions warmed up.

Prehistory of the North dismisses any assumptions that the colonization of the steppe-tundra was a simple process. Hoffecker reveals the complex dynamics of living in cold environments, where settlement ebbs and flows like the movements of a giant pump driven by climatic shifts. He describes the sophisticated hunter-gatherer cultures of the Don Valley and the Ukraine with their sophisticated dome-shaped houses that literally hugged the ground. He shows how extreme cold played an important role in the first human settlement of extreme northeast Siberia and the Americas. For the first time, we have an authoritative synthesis of what little is known about far northeast Asia at the time when the Bering Land Bridge joined Siberia and Alaska. From it, the case for a late settlement of the Americas, at the earliest by 19,000 years ago, becomes increasingly compelling.

The Cro-Magnon stereotype lingers in another sense, for all too often we archaeologists tend to ignore what happened in colder latitudes during global warming after the Ice Age. We learn of late Ice Age hunters in Eastern Europe grappling with newly forested environments. This book shows how the efficient cold-climate adaptations of the late Ice Age shifted focus in postglacial times, changing, adapting, in the north becoming more focused on the ocean and shoreline as sea levels rose, especially in areas where natural upwelling from the sea bottom produced a rich bounty for fisherfolk.

One cannot write a book like this without a sophisticated perspective on ancient landscapes. Hoffecker discusses such arcane topics as latitudinal zonation, which played an important role in isolating northern peoples in later millennia. He dwells expertly on the maritime arctic societies that flourished in the far north of both Europe and North America, many of which are little known outside the narrow coterie of northern specialists. We learn about the harsh realities of people in what is now Greenland with technology so basic that they may have spent the winters in a form of hibernation. And he shows how the isolation of the far north broke down during the twentieth century, disrupting thousands of years of fine-tuned adaptation to the some of the most severe environments on earth.

Prehistory of the North is an important book because it summarizes some of the least known archaeology on earth without regard to national boundaries, or even continents. The author has embraced an extraordinary range of obscure sources from many disciplines, even cold-weather medicine, and melded them into a provocative narrative that will serve as a stepping stone for new research. And, in this era of increasingly specialized research, it's a pleasure to read a bold, literate, and wide-ranging account of a huge subject written by a scholar who has the courage to set forth his ideas, well aware of the probability, horror of horrors, that some of them will be proven wrong in coming years. We need more archaeologists like John Hoffecker, who realize that boldness and a broad vision can advance scientific knowledge in ways unimaginable with more specialized inquiries. Herein lies the importance of a book that promises to influence generations of students. This really is a synthesis that should be on every archaeologist's reading list.

Brian Fagan

Preface

This book tells the story of how humans—who evolved in the tropics—came to occupy the colder regions of the earth. To my knowledge, it is a story that has not been told before, at least in book form. The reason probably lies in the fact that archaeologists tend to specialize in the study of particular places and time periods, and this strain of prehistory cuts across many of both. An unusual combination of circumstances exposed the author to many of these places and time periods, creating the basis for the book.

Examined as a whole, the settlement of higher latitudes presents a complicated picture. Early humans did not simply drift northward as their ability to cope with cooler climates gradually evolved. Their occupation of places like Europe and Northern Asia, and the later movement of modern humans into the Arctic and the New World, was achieved in relatively rapid bursts of expansion.

These episodes of expansion were often longitudinal as well as latitudinal, reflecting historical factors and the geography of climate. And their causes were varied, reflecting both anatomical and behavioral adaptations to cold environments. Only modern humans with spoken language—using their ability to structure their environment and social relations in complex ways—overcame the extreme conditions of the Arctic.

Acknowledging the help of others with a book that draws on experience spanning most of my professional career is not a simple task. I would like to thank at least some of the principal authors of that experience in chronological order. My primary exposure to the study of late Middle and early Late Pleistocene sites in Europe (that is, the Neanderthals and their predecessors) took place in the Northern Caucasus (1991–2000), and I thank G. F. Baryshnikov, L. V. Golovanova, and V. B. Doronichev for inviting me to help with the analysis of *Treugol'naya Cave, Il'skaya, Mezmaiskaya Cave,* and other sites in this part of the world.

More recently (2001–2003), it has been my great privilege to work with M. V. Anikovich, A. A. Sinitsyn, V. V. Popov, and others on the earliest modern human sites in Eastern Europe at *Kostenki*. And both in the Northern Caucasus and at Kostenki, I have learned much from my collaboration with G. M. Levkovskaya.

In broader terms, I owe much of my understanding of the transition to modern humans to Richard G. Klein, who was my Ph.D. advisor at the

University of Chicago. More than anyone else, Professor Klein has shed light on this most critical event in human evolution, which I have sought to explain from the perspective of the higher latitudes.

Although I have never worked in Siberia, I have benefited greatly from the insights and knowledge of colleagues who have, especially William Roger Powers, who was my M.A. advisor at the University of Alaska. A teacher, collaborator, and friend, Professor Powers died in September 2003 while I was revising the draft of this book, which is dedicated to his memory. I am also grateful to Ted Goebel and Vladimir Pitul'ko, and would like to thank Z. A. Abramova, who kindly showed me artifacts from the remarkable site of *Dvuglazka* in 1986.

For many years (1977–1987) I worked with both Professor Powers and Ted Goebel in the Nenana Valley, which contains some of the earliest known sites in Alaska and Beringia. Both at the outset of this research and in later years, all of us learned much about Beringian life and landscape from R. Dale Guthrie.

Since 1998 I have been studying late prehistoric remains on the arctic coast of Alaska—primarily at *Uivvaq* (Cape Lisburne)—and I would like to thank my collaborators Owen K. Mason, Georgeanne L. Reynolds, Diane K. Hanson, Claire Alix, and Karlene Leeper. I am especially grateful to Scott A. Elias (University of London), who researched paleoclimate history at Uivvaq through the study of insect remains and taught me much about past environments.

In recent years I have become acquainted with an entirely different set of historic remains in the North American Arctic and Subarctic. In a series of historic preservation projects undertaken for the U.S. Department of Defense during 1996–2001, I collected information on many military facilities of the Cold War era in Alaska and Greenland. These studies sparked an interest in technological innovation and change, which is reflected in this book, and I am grateful to my collaborators, especially Gary Kaszynski, Mandy Whorton, and Casey Buechler.

A number of colleagues at the University of Colorado have helped me to understand various issues addressed in this book, including especially Alan Taylor and Paola Villa. At the Institute of Arctic and Alpine Research, these colleagues also include John T. Andrews, E. James Dixon, John T. Hollin, and Astrid E. J. Ogilvie.

Many of the sites at which I have undertaken field research are mentioned in the book, and I acknowledge the various institutions that funded this and other related research, including the Leakey Foundation, National Science Foundation, Alaska Division of Parks, National Geographic Society, International Research and Exchanges Board, National Academy of Sciences, and the U.S. Department of Defense.

The writing of this book owes much to the science editor at Rutgers University Press—Audra J. Wolfe—who enthusiastically embraced the project and helped move it forward. I am also grateful to various scholars who reviewed and commented on individual chapters of the draft, including G. Richard Scott (chapter 2), Paola Villa (chapter 3), Philip G. Chase (chapter 4), Ted Goebel (chapter 6), and Owen K. Mason (chapter 7).

My thanks also to Elizabeth Gilbert, who edited the final manuscript. The maps were prepared by my wife, Lilian K. Takahashi, and the line drawings were done by my son, Ian Torao Hoffecker.

February 2004

A PREHISTORY OF THE NORTH

CHAPTER 1

Vikings in the Arctic

In AD 1000, the Earth was experiencing an episode of climate warming similar to that of the present day. Temperatures in many parts of the world seem to have risen by at least two or three degrees Fahrenheit. Although the scale of this "global warming" may seem small, its effects on human societies were profound. In Europe, several centuries of long hot summers led to an almost unbroken string of good harvests, and both urban and rural populations began to grow. These centuries are known as the Medieval Warm Period.[1]

One of the more dramatic consequences of the Medieval Warm Period was the expansion of Viking settlements in the North Atlantic. From their Icelandic base (established in AD 870), the Norse people began to move west and north to Greenland, Canada, and eventually above the Arctic Circle.

The discovery of a green stone inscribed with runic characters near Upernavik in northwestern Greenland indicates that a small party of Vikings ventured as far as 73° North (probably in the late thirteenth century). The inscription lists the names of three Norsemen and mentions the construction of a rock cairn, which was still present when the runestone was discovered in 1824. The Vikings had reached a point only 1,200 miles (1,900 km) from the North Pole. Their artifacts have been found even further north in Greenland and Ellesmere Island, but it is unclear who brought them to these locations.[2]

As they established settlements along the coast of Greenland and probed further into northern Canada and the Arctic, the Norse encountered native peoples of the New World. It was the first meeting of Europeans and aboriginal Americans. Although the Vikings were inclined to lump all of these peoples into the pejorative category of *skraeling*, they comprised a diverse array of groups. In the southern part of the Norsemen's range (for example, Newfoundland), they found Algonquian-speaking Indians. Both the Norse sagas and archaeological evidence suggest that interactions between Vikings and Indians were relatively limited.[3]

Farther north, the Vikings encountered very different sorts of people. In some places, such as northern Labrador and Baffin Island, they almost certainly met up with the last of the Paleo-Eskimo population (known to archaeologists as "Late Dorset"). These people were descendants of the earliest inhabitants of Greenland and the Canadian Arctic. Although capable

1

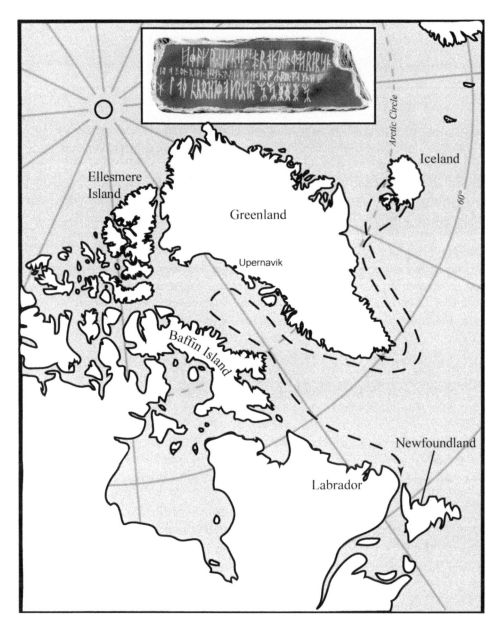

Figure 1.1. Map of Viking exploration and settlement in the North Atlantic during the Medieval Warm Period (AD 1100–1300). Inset: Runestone discovered near Upernavik in Greenland marking Viking presence at latitude 73° North.

hunters of walrus and polar bear and fully adjusted to arctic conditions, the Dorset possessed a comparatively primitive technology. Among other things, they lacked large boats and bows and arrows. Despite the warming climate, their settlement began to shrink after AD 1000, perhaps in response to other people in the region. Evidence for contact with the Norse is scarce, and it is widely assumed that the Dorset avoided the latter as much as possible.[4]

The native Americans with whom the Vikings interacted most extensively were the ancestors of the modern Inuit or Eskimo. The Inuit were themselves newcomers to the region, having spread eastward from Alaska after AD 1000. In fact, their movement into the Canadian Arctic and Greenland was probably facilitated by the same warming climates that had encouraged the Vikings to come north.

The Inuit were a formidable people with a tradition of warfare. They hunted bowhead whales in large boats *(umiaks)* and moved swiftly across the landscape in dogsleds. Their hunting technology and weaponry were highly sophisticated and included mechanical harpoons and recurved bows. Their winter clothing, which was assembled from more than a hundred components, provided effective protection from extreme cold.[5]

Inuit settlements were established on Ellesmere Island, northern Greenland, and other parts of the region by AD 1300. Inuit oral tradition, Norse sagas, and the evidence of archaeology suggest both trade and warfare occurred with the Vikings during the following two centuries. In many respects, this was the first serious contest between Europeans and native Americans.

Unlike later conflicts between the two peoples, the Vikings probably did not enjoy major advantages in terms of technology or numbers. Their boats were larger and powered by sail, and they made use of iron weapons and armor (which the native Americans sometimes tried to obtain through trade). However, the Norse settlers in Greenland were not the heavily armed Viking raiders of European legend, and local sources of iron were unknown. Most important, the Vikings lacked firearms. Written and oral history sources suggest that the Inuit may have been equally—if not more—aggressive, and that at times they assembled large numbers of people for attacks on the Norse.[6]

Although the victory is not widely appreciated, it is apparent that native Americans won their first contest with European invaders. By AD 1500, the Norse settlements in Greenland and elsewhere in the New World had been abandoned. The Dorset people had also disappeared by this time, and the Inuit inherited all of the arctic—and some of the subarctic—regions of the New World.

The reason for the retreat of the Vikings from these regions has been the subject of much debate. Economic competition and warfare with the Inuit seem likely to have been factors, along with declining trade and the isolation of the settlers from the larger Norse population. The primary cause, however, probably lies in the return of colder climates that heralded the beginning of the "Little Ice Age" in AD 1450–1500. Falling temperatures were almost certainly the reason for the economic decline that took place at this time and the reduction in population that followed. Conflict with the Inuit probably exacerbated Norse problems, but did not create them.[7]

The real obstacle to Viking survival in the north was their inability to adapt to colder climates during the 1400s. The Inuit were also forced to make adjustments to their way of life at this time (for example, increased focus on seal hunting), but they seem to have accomplished this without major trauma and within the larger context of their existing adaptation.

Isotopic analyses of the skeletal remains of Greenland Vikings, combined with the study of food remains from their settlements, indicates that they gradually adopted a diet based more heavily on marine foods (and less on livestock).[8] However, they never abandoned the fundamental traditions of a society and culture derived from medieval Europe. Dressed in woolen clothing, they were still struggling to maintain their farming estates as arctic climates descended on southern Greenland.[9]

The Settlement of Cold Environments

Although the Vikings could not know it, their movement north during the Medieval Warm Period of AD 1000–1400 represented a pattern that had occurred many times before in the human past. Throughout prehistory and history, peoples have shifted their range northward in response to improved climates. Conversely, they have sometimes retreated from higher latitudes during phases of colder climate.

The initial movement of early humans above latitude 45° North roughly half a million years ago may have been largely a consequence of warmer climates. The peak of the last major glacial advance 24,000 years ago seems to have forced modern humans to abandon large areas of northern Eurasia. And rising temperatures in Siberia toward the end of the Ice Age (roughly 16,000 years ago) encouraged people to occupy the Bering Land Bridge and enter the New World.[10] There are many other less spectacular examples from later prehistory and historic times.

The pattern of northward movement during episodes of warmer climate is one aspect of the human settlement of northern latitudes. The same pattern may be found among plants and animals as they shift their range in response to changes in temperature and moisture. During the warm interval that prevailed between 7,000 and 4,000 years ago, boreal forest vegetation spread northward beyond its current limit. At the time of the last major glacial advance 24,000 years ago, many animals now confined to the arctic tundra (for example, polar fox, musk ox) extended their range hundreds of miles southward to places like southern Ukraine.[11] In these cases, organisms have simply followed the shifting boundaries of their environment without developing significant new adaptations. Accordingly, these organisms were forced to retreat when climate trends reversed direction and the boundaries shifted back.

The Inuit represent a different aspect of human settlement in higher latitudes and colder regions. Unlike the Vikings, they had developed a wide range of adaptations to their arctic maritime environment. In addition to their highly sophisticated and specialized technology, the Inuit had developed organizational strategies for coping with the challenges of this environment. They had also evolved morphological traits and physiological responses that helped them conserve body heat and avoid cold injury. They were the supreme arctic specialists, and probably overwhelmed the less efficient Dorset people in addition to competing successfully with the Vikings.

Like the latitudinal shifts of peoples during periods of climate change, the development of specialized adaptations to northern environments also has many parallels among other living organisms. Most plants and animals that live at higher latitudes represent taxa specifically adapted to conditions in those latitudes. These organisms have diverged from their ancestral forms, evolving new features in response to the lower temperatures, increased seasonality, reduced sunlight, and other aspects of northern environments.

During the course of their evolution, humans produced at least one specialized northern variant in the form of the Neanderthals, who diverged roughly half a million years ago from the southern population and eventually developed a variety of cold adaptations. Since the appearance and spread of modern humans more than 50,000 years ago, however, people have adapted to higher-latitude environments primarily through cultural means (although many—like the Inuit—developed some physical adaptations without becoming genetically isolated from other modern humans).

A Prehistory of the North

The settlement of northern lands and coastlines has been a major theme in history and prehistory. Humans evolved in the tropics and for the most part have never really "belonged" in cold places. Their occupation of the latter has always required some change—either change in their abilities to cope with conditions in these places or changes in the places themselves, or some combination of the two.

A variety of problems confront a tropical plant or animal attempting to spread northward into cooler environments. The most obvious of these is lower temperature—especially during winter months.[12] A number of other challenges, however, may be equally or even more important than temperature.

Although some cold marine environments are very rich, cooler environments tend to be less productive than those of the equatorial zone. This is largely because cooler terrestrial environments are generally drier, and the reduced moisture limits plant growth, and this in turn supports less animal life.[13] Cooler environments also tend to be increasingly seasonal, and the variations in temperature and moisture may cause sharp fluctuations in resources for some organisms. Seasonality reaches an extreme level in arctic continental settings, where the difference between the mean January and July temperatures is often more than 100° F (38° C).[14]

This book is about the settlement of cold places. It is primarily an attempt to explain how humans achieved each successive advance into the middle and higher latitudes. Although the emphasis falls on those advances that entailed new adaptations to cold environments, changing climates seem to have played a critical role in the process of developing these adaptations.

Despite the fact that higher latitudes and cooler environments exist in the southern hemisphere, the focus here is entirely on the North. There is little land below latitude 30° South (that is, the limit of the tropical zone) with the exception of southern South America and Antarctica. South America was inaccessible to humans until the end of the Ice Age, while the existence of Antarctica was unknown until late historic times. The settlement of the colder parts of the Earth is therefore a prehistory of the North.

A review of the human fossil and archaeological record over the past 5 million years (that is, since the first appearance of the human family) reveals that the settlement of higher latitudes and colder environments did not occur as a result of the gradual northward drift of populations and cultures. Instead, each major advance seems to have taken place relatively quickly, as climate change or new adaptations suddenly opened new regions and habitats for occupation.

Moreover, because of the influence of oceans and continents on terrestrial climate, many of these advances were longitudinal rather than latitudinal—movements along a climate gradient that ran from east to west as much as from north to south. This is particularly evident in northern Eurasia, where the "oceanic effect" of the North Atlantic brings milder climates to Western Europe, while colder and drier conditions prevail in Eastern Europe and Siberia.[15]

Five major stages may be defined in the human settlement of the North.
Stage 1: Occupation of the Middle Latitudes. Between roughly 1.8 and 0.8 mil-

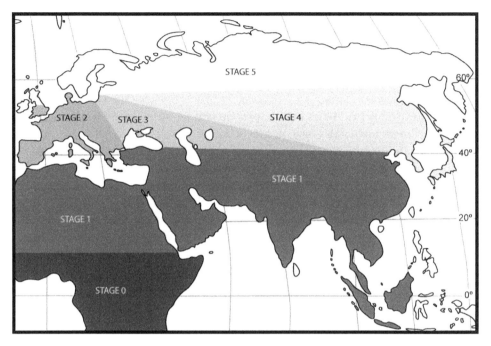

Figure 1.2. Map of Africa and Eurasia illustrating broad patterns of settlement in middle and higher latitudes by humans between 1.8 million and 7,000 years ago.

lion years ago, early humans expanded out of their tropical African base and colonized Eurasia as far as latitude 41°–42° North. This stage is primarily associated with *Homo erectus* and changes in anatomy and behavior that allowed humans to forage across open and comparatively dry landscapes. Although perhaps rarely—if ever—exposed to subfreezing temperatures, *Homo erectus* populations coped with less productive and more seasonal environments than their predecessors. Their adaptations to these environments set the stage for subsequent expansion into higher latitudes.

Stage 2: Colonization of Western Europe. Between at least 500,000 years ago and up to roughly 250,000 years ago, humans (most of whom may be assigned to the taxon *Homo heidelbergensis*) occupied the continent of Europe as far east as the Danube Basin. In Britain, sites in this time range are found as far as latitude 52° North. With the possible exception of controlled fire, obvious adaptations to cold are lacking in the human fossils and archaeological sites of this interval. The initial colonization of Europe may have been largely an opportunistic expansion into the warmest parts of northern Eurasia—previously blocked by factors other than cold climate. Alternatively, some cold adaptations may remain concealed by the poverty of *Homo heidelbergensis* fossils and their archaeological record.

Stage 3: The Neanderthals. The Neanderthals *(Homo neanderthalensis)* evolved gradually in Western Europe and expanded eastward into colder and drier parts of northern Eurasia by at least 130,000 years ago. They became the first humans to occupy the central East European Plain and southwestern Siberia. Unlike their predecessors in Western Europe, the Neanderthals exhibit a suite of anatomical and behavioral adaptations to cold environments. A diet high in protein and fat—obtained from the hunting of large mammals—was of critical importance.

Despite the Neanderthals' special cold-adapted traits, their range of climate tolerance was limited compared with that of modern humans. They probably were unable to cope with average winter temperatures much below 0° F (−17° C) and were generally restricted to wooded terrain.

Stage 4: Dispersal of Modern Humans. Between 45,000 and 20,000 years ago, modern humans—originally derived from Africa—expanded into habitats and regions never occupied by early humans. The regions included Siberia as far as latitude 60° North (and sometimes farther, at least on a seasonal basis). Modern humans' success was chiefly due to an ability to develop complex and innovative technology (for example, insulated clothing, artificial shelters), some of which was essential to survival at higher latitudes during the middle of the Last Glacial period—where mean winter temperatures often fell below −5° F (−20° C). Flexible organization, however, may have been an important factor in sustaining a population in very cold and dry habitats, where resources were widely scattered. Both novel technology and flexible organization were probably related to language and the use of symbols.

The modern humans who invaded northern Eurasia 45,000 years ago retained the warm-climate anatomy of their recent African ancestors. This may have kept them out of the Arctic (that is, above latitude 66° North) and forced them to abandon the colder parts of northern Eurasia (including most of Siberia) as the Last Glacial reached its cold maximum about 24,000 years ago.

Stage 5: Modern Humans in the Arctic. The final stage may be divided into two substages. The initial occupation of arctic environments took place between roughly 19,000 and 7,000 years ago, as modern humans reoccupied parts of northern Eurasia abandoned during the peak of the Last Glacial. Several factors—including postglacial warming and some anatomical cold adaptations—may have triggered this event. Milder climates opened the door to northeast Asia and the Bering Land Bridge, and humans crossed into the Americas for the first time. After 7,000 years ago, humans expanded into deglaciated areas of Canada and other previously uninhabited regions of the Arctic. Much of their success was based on technological innovation (for example, large boats, toggle-head harpoons) that facilitated a robust maritime economy.

The industrial civilization that arose initially in Western Europe during AD 1250–1700 eventually colonized much of the world, but was slow to move into the Arctic. Following initial explorations, there were limited efforts to exploit marine mammals and mineral resources. Today, few cities lie above 60° North and no major urban centers are found above the Arctic Circle.

CHAPTER 2

Out of Africa

Forest Man

Our closest relatives are today confined to forests located near the equator. Gorillas and chimpanzees inhabit a belt of tropical forest and woodland that stretches across western and central Africa. Orangutans, somewhat less closely related to humans than the African apes, are found in similar environments on the islands of Sumatra and Borneo in the Malay archipelago.[1] In the Malay language, *orangutan* means "forest man."

Our proximity to the living great apes was first perceived through the study of comparative anatomy. In 1698 the anatomist Edward Tyson acquired the body of a chimpanzee brought to London by an English sailor on a ship from Angola. Tyson undertook a careful study of his specimen and drew comparisons with the anatomy of humans and monkeys. He concluded that the chimpanzee exhibited a greater resemblance to a human being.[2] A few decades later, the Swedish naturalist Linnaeus published his classification of living plants and animals, placing both humans and apes into the genus *Homo*.[3]

Chimpanzees are somewhat smaller than humans, and the average weight for males is about 100 lbs (45 kg) and for females about 88 lbs (almost 90 percent of male body size). Gorillas are larger and more sexually dimorphic (that is, there is a greater difference between the sexes). Adult males weigh over 300 lbs (136 kg) and females are only about half the size of males. Orangutans are closer in body size to chimpanzees, but share the pronounced sexual dimorphism of the gorillas. Unlike living humans, chimpanzees and gorillas are covered with a thick coat of hair. Ape brains, though large by most standards, are significantly smaller than those of humans relative to overall body size (roughly one-third of human brain volume).[4]

One of the characteristics that sets the African apes apart from humans, as well as from orangutans, is their unique mode of locomotion, which is known as *knuckle-walking*. On the ground, chimpanzees and gorillas walk with their feet flat but their hands curled to touch the surface with the knuckles. Both chimpanzees and orangutans spend much of their time in the trees. Adult gorillas rarely climb trees because of their size.[5]

The behavioral similarities between apes and humans were also noted long ago by various naturalists. As early as 1844, wild chimpanzees were ob-

Figure 2.1. Chimpanzee using stick to fish for termites.

served using stones to break open tough fruit kernels. Some years later Charles Darwin described watching a young orangutan employ a stick as a lever.[6] In 1963 Jane Goodall nullified the unique status of humans as "toolmaking animals" when she reported wild chimpanzees manufacturing simple tools from plant materials. These included sticks stripped of twigs for fishing termites out of their nests, and leaves chewed for use as sponges.[7] Gorillas have rarely been observed making or using tools.

Another chimpanzee behavior of special interest to students of human evolution is hunting. Wild chimpanzees have been observed to hunt monkeys and small antelopes, and they sometimes hunt in cooperative groups. After a hunt, the meat may be shared with others. Nevertheless, meat represents a small part of the chimpanzee diet, which is primarily composed of ripe fruits and berries, along with some seeds, nuts, leaves, and insects. Orangutans also consume large quantities of fruit and some insects, while gorillas feed chiefly on leaves and plant shoots obtained on the forest floor.[8]

Both chimpanzees and gorillas possess a relatively thin coat of enamel on their teeth, which reflects the chewing of soft foods that cause little abrasion and wear on the tooth surface. By contrast, human teeth have a comparatively thick coat of enamel, indicating a dietary history of harder and more abrasive foods. Orangutans' tooth enamel is of intermediate thickness, and apparently reflects their consumption of tougher fruits and other items.[9]

It was molecular research that revealed how close the genetic and evolutionary relationship actually is between humans and the African apes. Immunological studies of blood proteins performed in the early twentieth century demonstrated—as expected—that apes were more closely related than Old World monkeys to humans.[10] In the 1960s these studies were resumed and expanded with new techniques, yielding startling results. The new molecular analyses indicated an extremely close relationship

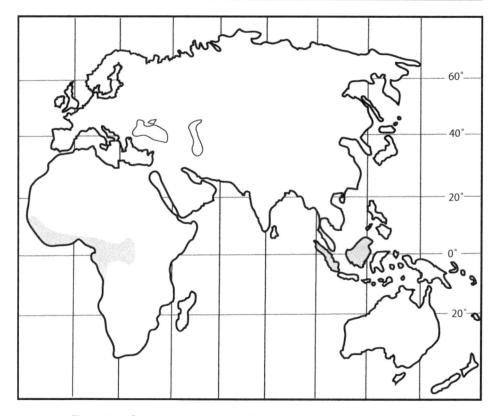

Figure 2.2. Geographic distribution of the living apes.

among humans and chimpanzees and gorillas. Some of these studies suggested that chimpanzees were more closely related to humans than they were to gorillas, though previously reported estimates of 98.5 percent shared DNA now appear to be too high.[11]

Despite the close evolutionary link between humans and living apes, it is readily apparent that in one respect they differ in the extreme. Both the African apes and the orangutans are restricted to a narrow geographic range below latitude 15° North. Chimpanzees are the most versatile, and some of them inhabit open woodlands, while gorillas and orangutans are entirely confined to tropical forest within a few hundred miles of the equator. The living apes inhabit a small world where the average temperature is roughly 80° F (25° C) and rarely varies more than a few degrees during the year. Moisture is generally abundant and plant productivity is high. Humans—who have achieved almost global terrestrial distribution within the last few thousand years—occupy desert, boreal forest, and tundra, having invaded some of the coldest, driest, and least productive environments on earth.

Epoch of Global Change

The molecular research of the 1960s ultimately forced a major reassessment of human evolution. Not only did the new evidence indicate an unexpectedly close genetic relationship between humans and African apes, but—when used to estimate the date of divergence between the two—the "biomolecular clock" suggested an evolutionary split less than 8 million years ago.[12] At the time of the publication of these findings, most anthropologists believed that humans had diverged from a forest-dwelling African ape roughly 15 million years ago.[13] Accordingly, the molecular evidence was largely dismissed or ignored until the late 1970s, when the fossil record was reevaluated.[14]

The dates provided by the biomolecular clock are now widely accepted as at least rough estimates of divergence times among the various living species. Both the molecular evidence and the comparative anatomy of fossils suggest that the ancestors of living orangutans diverged from the gorilla-chimpanzee-human lineage about 12 million years ago.[15] But a major gap in the African fossil record between 10 and 5 million years ago—combined with uncertainties about interpreting the molecular data—continues to obscure the evolutionary relationships among humans, chimps, and gorillas.

At present it is unknown if the last common ancestor of all three inhabited forest or more open landscape. Analysis of tooth enamel in living apes and humans suggests that the thicker enamel of the latter is the ancestral condition, and knuckle-walking also appears to represent a specialized adaptation.[16] Humans are descended from African apes only in a broad sense. Their last common ancestor probably differed in many respects from the living representatives of both groups and may have occupied an open woodland environment.

All of these developments took place during the Miocene epoch, which lasted from 23.5 to 5.2 million years ago. The Miocene was a period of major change in earth history that began with significantly warmer temperatures and more extensive forest environments in Africa and Eurasia. At that time, the two land masses were separated by water. Roughly 18–17 million years ago, they collided and many animals spread across the new land connection from Africa into Eurasia. Climates became cooler and drier, and grassland and woodland expanded as the forests shrank.[17]

Both Old World monkeys and *hominoids*—a group that includes all living and fossil apes and humans—evolved in Africa during the early Miocene. The monkeys, which were generally adapted to more open environments, were less widespread and diverse at this time. In addition to the well-known primitive ape *Proconsul,* several hominoid genera were present

14   A PREHISTORY OF THE NORTH

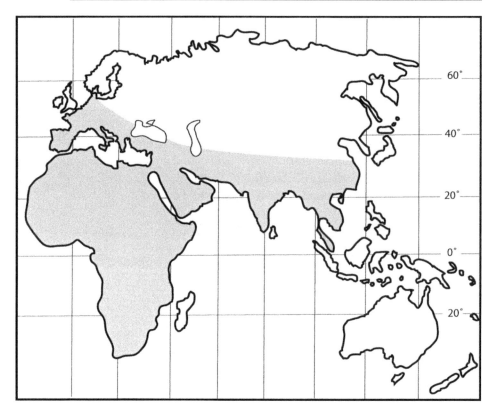

Figure 2.3. Distribution of apes during the middle Miocene (17–7 million years ago).

before 18 million years ago. They varied in terms of size, locomotion, and dietary adaptations, but all of the early Miocene forms known to date were restricted to the forests and woodlands of tropical east Africa.[18]

After the collision of the African and Eurasian continents, Old World monkeys and hominoids expanded their range across southern Asia and into parts of Europe. The mid-Miocene distribution of hominoids was similar to that of humans roughly half a million years ago (see chapter 3), and the period between 17 and 7 million years ago became a golden age for apes in terms of their numbers and variety. As many as eleven different species of *Pliopithecus* inhabited forests from central Europe to southern China between 16 and 11 million years ago. Smaller than the modern African apes, they were arboreal quadrupeds like the living monkeys. Also present was *Dryopithecus*, which evolved a more apelike body and elongated forelimbs. These taxa ranged as far as 50° North in Europe, where subtropical climates prevailed at the time.[19]

Several species of *Sivapithecus* occupied drier and more open woodland environments in both Europe and Asia. The sivapithecines possessed thick enamel on their teeth and consumed a diet of more resistant foods. They are widely believed ancestral to modern orangutans. More exotic hominoids without direct links to living apes appeared during the later Miocene. *Oreopithecus* inhabited swampy forests in Italy roughly 9–8 million years ago. In part because of its unusual dental pattern, it is normally placed in a separate and extinct family. Ancient lake and swamp deposits in southern China yielded *Lufengpithecus,* which exhibits a unique and even bizarre-looking skull with a concave forehead. The massive *Gigantopithecus*—largest of all known primates—also appeared in the Far East at the end of the Miocene.[20]

Roughly 7 million years ago, another sharp decline in temperatures took place. As polar ice caps grew and global sea level fell, climates became cooler and drier across much of Eurasia and northern Africa. By 6 million years ago, the Mediterranean Sea was a dry basin. Forest environments were further reduced while grasslands increased. As Old World monkeys expanded and diversified, the hominoids went into a decline from which they never recovered. They became completely extinct in Europe and less common in southern Asia.[21]

Human Origins

Humans were once thought to have evolved as part of the great expansion and diversification of hominoids during the mid-Miocene. As already noted, the biomolecular research that began in the 1960s eventually forced a reassessment of this view. It is now clear that the earliest humans appeared at the close of the Miocene, when hominoids were suffering widespread extinction and range contraction. Humans became one of the few surviving hominoid lineages by evolving a highly unusual feature that allowed them to occupy an ecological niche in a tropical woodland environment.

Although the details of their emergence remain hidden by the scarcity of African fossils between 10 and 5 million years ago, humans are present by the end of this interval (early Pliocene epoch). The oldest known representatives of the *hominid* family are the australopithecines, who inhabited Africa until roughly 1 million years ago. At present, the earliest remains—assigned to *Ardipithecus ramidus*—are found in East Africa and are dated to slightly more than 5 million years ago. Though confined to tropical latitudes between 16° North and 27° South, the australopithecines were relatively diverse, and up to three genera and at least seven species are recognized by most anthropologists.[22]

Had they survived to the present day, the australopithecines would most likely be regarded as apes owing to the small size of their brains, which varied between 400 and 550 cc (comparable to the living apes in proportion to

16   A PREHISTORY OF THE NORTH

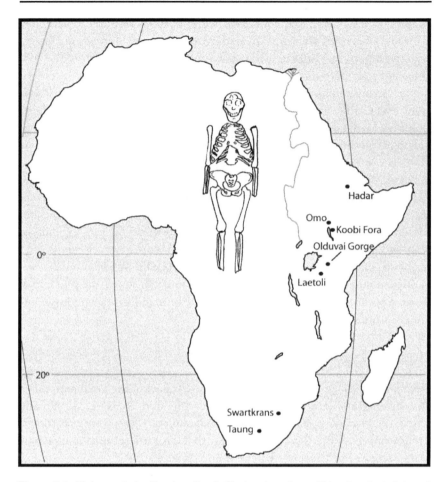

Figure 2.4. Major australopithecine sites in East and southern Africa. Inset: skeleton of *Australopithecus afarensis* ("Lucy").

overall body size). They retained many other apelike features, and their geographic range was almost as limited as that of the modern chimpanzee. However, the australopithecines had evolved a mode of locomotion—walking upright on their hindlimbs—that set them apart from not only the African apes but all other living primates and most mammals. It was the development of *bipedalism* that moved humans onto their fateful evolutionary track. Together with the later appearance of language, it remains the most important event in human evolution.

Despite its rarity among mammals, bipedal locomotion had clear roots in the upright posture of the primitive apes and lower primates. Although the transition required a host of interrelated changes in morphology and function that affected much of the skeleton, it was less drastic than it might seem. The lower spine became curved and the pelvis became shorter and wider. The position of the skull shifted to rest vertically above the trunk. The lower limbs became longer; the surface areas for the joints became larger. The feet assumed a platform structure with a nonopposable big toe.[23]

Much speculation and debate have surrounded the origin of bipedalism, although there is consensus that it was tied to the expansion of open habitat toward the end of the Miocene. As the distribution and density of trees declined, hominoid food sources probably became more widely dispersed, and some of the late Miocene apes could have been forced to spend more time on the ground. Bipedalism—which is a highly energy-efficient mode of locomotion—may have evolved among one or more of them as a response to the increased need for movement across open habitat.[24] It may have had other advantages that would help explain its appearance at this time, such as enhancing the ability to carry food and/or offspring and to see over tall grasses and shrubs while foraging on the ground.[25]

Bipedalism is also related to the making and using of tools, but the causal relationship between the two is difficult to untangle. Bipedalism clearly freed the forelimbs for tool production and—perhaps even more important—for weapon use while walking or running. However, none of the tools found in Africa that date to more than a million years in age can be firmly associated with any of the australopithecines. It is widely assumed that they made and used tools similar to those found among the living great apes, but these remain archaeologically invisible (primarily because the australopithecines did not habitually occupy and leave debris at specific locations). Chimpanzees make and use their tools without bipedalism, and some anthropologists believe that the tool-making advantages of the latter came later.[26]

Although the australopithecines were bipedal, they retained some ape-like features that suggest they were still spending much time in the trees. Like the great apes, their arms were longer than their legs. The arm socket (the glenoid cavity of the scapula) appears to have been positioned close to the head, facilitating upward movement of the forelimbs for climbing and suspension. *Australopithecus* fingers and toes were curved (the toes were elongated), and especially suited for gripping branches. Finally, the structure of the inner ear—which plays a critical role in maintaining balance during bipedal locomotion—was more similar to that of apes than modern humans.[27] Presumably trees were an important source of food and a refuge from predators.

Other clues to the diet and foraging practices of the australopithecines are provided by their teeth and jaws. The former were typically covered with a thick coat of enamel, indicating a diet of comparatively hard foods relative to the living great apes. The cheek teeth (premolars and molars) were rather large, while their front teeth were smaller. The later australopithecines evolved a powerful chewing complex with even larger cheek teeth and massive jaws for lateral grinding of food items.[28] The younger, more robust forms are often assigned to a separate genus *(Paranthropus)* and were contemporaneous with early representatives of the genus *Homo*.

The australopithecines never expanded their geographic range beyond the tropical zone and never colonized Eurasia. Their remains are found no further than 16° North, and although this latitude encompasses the southernmost portion of the Arabian Peninsula, it is well below the point where the latter provides access to the Eurasian landmass (26° North at the Strait of Hormuz). The limits of the australopithecine range were probably related to their heavy dependence on trees for food and protection, although perhaps their diet (which apparently included little or no meat) was a factor.[29]

Early *Homo*

Between 1960 and 1963 Louis and Mary Leakey recovered the bones and teeth of a previously unknown hominid from *Olduvai Gorge* in Tanzania. Prior to that time, the site had yielded only the remains of a robust australopithecine. Together with two other colleagues, the Leakeys subsequently announced the discovery of the oldest representative of our genus *(Homo habilis)*. The announcement proved controversial, and the status of these fossils—along with others assigned to the same or closely related species—has remained controversial to the present day. Much of the problem lies in the small size and pronounced variability of the sample of early *Homo* remains.[30]

Since 1960 bones and teeth assigned to *Homo habilis* (or other early *Homo* species) have been recovered from a number of east African localities and dated to between 2.5 and 1.6 million years ago. Their appearance in the fossil record marks several important new developments in human evolution. Early *Homo* exhibits a significant increase in brain size (both in absolute terms and relative to overall body size). Reflecting the wide variability in the sample, estimated brain volume ranges between 510 and 750 cc. The shape of the brain case was altered by the expansion of the temporal and parietal regions. For the first time, hominids exceeded the general size level of the ape brain.[31]

Although the teeth and jaws were reduced in comparison with their predecessors, early *Homo* anatomy remained very similar to that of the australopithecines in other respects. Limb bones from Olduvai Gorge and *Koobi Fora* (Kenya) indicate that the arms were still long relative to the legs. Overall body size was small and differences between the sexes—sexual dimorphism—seem to have been comparable to those of the living great apes. *Homo habilis* may have continued to spend time in the trees and live in social groups similar to those of the australopithecines.[32]

As in the case of bipedalism, the evolution of larger brains and tool-making abilities in early *Homo* has been explained as a response to the shrinkage of tree cover. After 3.0 million years ago, many African woodland taxa became extinct, while savanna-dwellers became more common. The earliest forms of *Homo* apparently evolved from one of the more gracile australopithecines (*Australopithecus africanus* or a similar species). At the same time, australopithecines became more robust, evolving the powerful chewing complex described earlier. This divergence was probably driven in part by competition between the two closely related hominids occupying similar ecological niches.[33]

The appearance of the large-brained early *Homo* coincides with evidence for important behavioral changes that have implications for human ecology. The first stone tools are also found in east Africa and dated to this interval (2.6–2.3 million years ago). Moreover, they are often found in association with animal bones that exhibit traces of tool damage. These concentrations of tools and animal bones represent the oldest known archaeological sites and indicate places where groups of early *Homo* camped and/or paused during their foraging rounds to perform various activities.[34]

The stone tools manufactured by early *Homo* are assigned to the *Oldowan* industry (Lower Paleolithic), which is present for at least a million years (2.6–1.6 million years ago). Oldowan tools are simple and confined to flaked cobbles and pebbles—often labeled "core tools"—or modified stone flakes. Few if any of them seem to reflect a design based on a mental template, and most of the tool types identified by archaeologists may be imposed on a continuum of variation by the modern human brain. Many of the pebble "tools" appear to have been created simply in order to obtain sharp flakes. Oldowan artifacts include hammerstones, choppers, proto-bifaces, flakes, and scrapers.[35]

Despite the simplicity of these tools, experiments demonstrate that their production is beyond the capacity of the living apes. In 1990 Nicholas Toth and several colleagues attempted to teach a young bonobo (or "pygmy chimp") named Kanzi how to make and use Oldowan stone tools. Although able to knock flakes off larger rocks (sometimes by hurling them against a hard floor) and use them for simple tasks, Kanzi was incapable of mastering the striking angles and other skills required to produce tools similar to those made by early *Homo*.[36]

Scanning electron microscope (SEM) analysis of animal bones in early *Homo* sites indicates that at least some of the Oldowan stone tools were used for smashing bone shafts—presumably to extract the marrow contained in

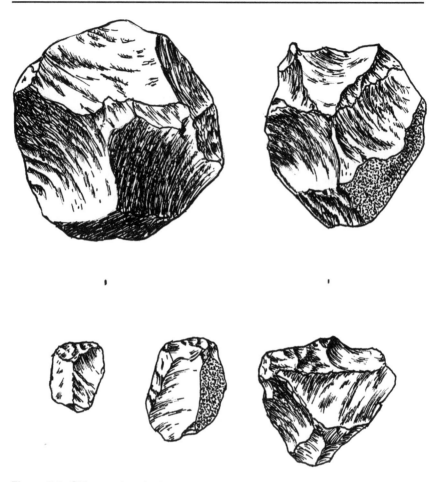

Figure 2.5. Oldowan stone tools.

their cavities—and stripping meat off their surfaces.[37] *Homo habilis* was apparently consuming more meat than either the great apes or the australopithecines, although it remains unclear whether this meat was obtained by hunting or merely by scavenging carcasses (which are relatively plentiful in tropical woodland and savanna). At the same time, comparative studies of tooth wear suggest that the early *Homo* diet did not differ fundamentally from that of australopithecines, chimps, or orangutans—the change in dental wear patterns came later.[38]

Early *Homo* sites are typically found along ancient lakeshores and watercourses. To some extent, this pattern may be influenced by rapid burial and preservation in such settings, but they are also locations where food, water, and shelter (in the form of trees) were likely to be found. The archaeologist Glynn Isaac suggested that the appearance of these sites marked an important shift in human ecology—the establishment of "home bases" to which food and other resources were transported. According to Isaac, home bases reflected both central-place foraging (bringing foods back to a central base) and the provisioning of females and young by meat-gathering and food-sharing adult males.[39]

Like the evolutionary status of *Homo habilis*, the home base hypothesis remains controversial. Although there is strong evidence for home bases and central-place foraging during the last few hundred thousand years, it is still not clear if they were early *Homo* adaptations. An alternative explanation for the concentrations of stone artifacts and animal bones is that they merely represent locations where both the raw materials for making tools and animal carcasses are found. The earliest human sites may be analogous to those of modern chimpanzees—temporary stopping places for food processing during the foraging round. Unlike chimpanzees, however, early *Homo* groups were processing a significant amount of meat at their stops.[40]

The continuing uncertainties about the morphology and behavior of *Homo habilis* make it difficult to define its ecological niche with any precision. Although the evidence for increased meat consumption—which seems to be related to both the larger brain and stone tools—suggests how this niche probably differed from that of the robust australopithecines, early *Homo* remains are found in the same places as those of *Paranthropus*. Like the latter, *Homo habilis*, despite its novel adaptations, seems to have been tied to tropical woodland. Hominid fossils older than 1.8 million years have yet to be found either above 16° North or outside of Africa.

Out of Africa

Between 1.8 and 0.8 million years ago (Early Pleistocene epoch), representatives of *Homo* expanded out of Africa and into southern Eurasia. Firmly dated archaeological sites in this time range are known from the Near East and Central Asia and from China and Southeast Asia. In the course of this expansion, humans dramatically increased both their geographic and their latitudinal range. Although earlier sites have yet to be found above 16° North, after 1.8 million years ago they are found as far as 41°–42° North, which appears to delineate a northernmost limit not exceeded until the beginning of the Middle Pleistocene (roughly 0.8 million years ago).[41]

The geographic expansion of *Homo* at this time may be equated with a highly significant expansion in habitat range. Humans were able to colonize northern Africa and southern Eurasia after 1.8 million years ago because they had evolved adaptations to environments where their predecessors could not survive. Although these environments lay outside the zone of cold climates—temperatures probably rarely fell below the freezing point—they

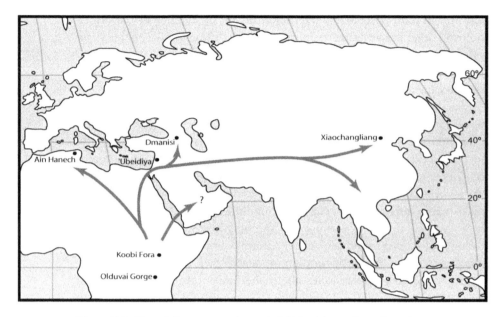

Figure 2.6. Map of *Homo* expansion out of Africa into southern Eurasia.

possessed two major features of cold environments. They reflected both the lower biological productivity and the pronounced seasonality of the higher latitudes.

Many of the areas occupied by *Homo* at this time exhibit low plant productivity in comparison with equatorial Africa. This is especially true for the drier regions of northern Africa and southwest Asia. Reduced plant productivity lowers the density of animal resources.[42] Thus humans had to adjust to habitats with fewer or more widely dispersed food resources. And because of the increased seasonality of areas at higher latitudes, they also had to adjust to major variations in climate and food availability during the year. By adapting to these conditions, they set the stage for the subsequent colonization of the North.

The pattern of habitat expansion after 1.8 million years ago is evident in tropical Africa. During 1.7–1.5 million years ago, sites in the dry upland areas of the Lake Turkana Basin (Kenya) were occupied by *Homo* for the first time. These sites were located in relatively open scrub and grassland habitat along ephemeral streams. By 1.5 million years ago, hominids were present on the Ethiopian plateau more than 7,000 ft (2,300 m) above modern sea level. Today this region is sometimes exposed to freezing or near freezing temperatures.[43]

Habitats occupied outside the tropical zone reveal even sharper contrasts with those of *Homo habilis* and the australopithecines. New dates from

Ain Hanech (Algeria) indicate that *Homo* had expanded into North Africa—reaching latitude 35° North—by 1.8 million years ago. And as early as 1.7 million years ago, humans were living (at least on a seasonal basis) at *Dmanisi* on the southern margin of the Caucasus Mountains (Republic of Georgia) at latitude 41° North. Although climates were somewhat warmer and drier than now, Dmanisi supported a subtropical or temperate woodland containing birch and pine with some steppic plants (for example, *Artemisia*). A mixture of typical African and Eurasian mammals roamed the area, including horse *(Equus stenonis),* gazelle *(Gazella borbonica),* and deer *(Dama nesti)*.[44]

Until recently, it appeared that the initial expansion out of Africa took place after 1.5 million years ago. This would have limited the earliest Eurasian hominids to representatives of *Homo erectus,* which exhibits some important anatomical differences with *Homo habilis*. Furthermore, by 1.5 million years ago, people in Africa were producing hand axes and other bifacial stone tools of the *Acheulean* industry (Lower Paleolithic), which reflected advances in their technological skills. Thus significant changes in morphology and behavior were thought to underlie the invasion of new habitats and regions.[45]

It is now apparent that the first hominids in Eurasia were making tools very similar to those of the Oldowan industry. Artifacts recovered at Dmanisi (roughly 1.7 million years ago) include simple core tools and flakes like those associated with *Homo habilis* at Olduvai Gorge.[46] Pebble and flake tools are also present in the oldest known East Asian sites, such as *Xiaochangliang* (Nihewan Basin, China), which is currently dated at 1.36 million years (and they continue to dominate industries in this part of the world until after 0.8 million years ago).[47] Hominid remains in Southeast Asia have yielded dates as early as 1.8–1.7 million years—also antedating the Acheulean—although these dates are disputed.[48]

Moreover, although the human skulls recovered at Dmanisi have been classified as *Homo erectus,* they reveal some striking similarities to *H. habilis*—especially the size of the brain (which ranges between 600 and 780 cc). Little information is currently available about the postcranial skeleton of the Dmanisi hominids.[49]

The possibility that the initial movement out of Africa might have taken place without major morphological or behavioral change is surprising, and this problem will require more investigation. In the meantime, it is clear that most African and Eurasian representatives of *Homo* during 1.8–0.8 million years ago differed from *H. habilis* in terms of size and a number of anatomical features. The changes are evident in African specimens dated to 1.8–1.7 million years ago, and they are present in East Asia by 1.0 million years ago at the latest.[50]

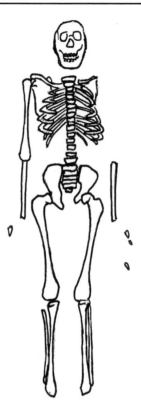

Figure 2.7. Nariokotome skeleton.

In late August 1984, Kamoya Kimeu made one of the most important of his many hominid fossil discoveries in Kenya. On a small hill near *Nariokotome* on the western margin of Lake Turkana, he found the first bone fragment of what later turned out to be a *Homo erectus* skeleton. Aside from the smaller bones of the hands and feet (missing), the skeleton was 66 percent complete. It belonged to a male with an estimated age at death of about twelve years.[51]

The Nariokotome skeleton was dated to 1.53 million years ago, and revealed much new information about humans at the time of the expansion out of Africa. Below the neck, *Homo erectus* was essentially modern in appearance and fully bipedal (lacking the long forelimbs and other features of the australopithecines and *Homo habilis* that suggested tree climbing). Presumably *Homo* was no longer tied to trees as a source of food and/or protection. There was increase in brain size relative to *H. habilis* (up to a mean of roughly 900 cc), but a corresponding increase in overall body size accounts for most or all of the larger cranium.[52]

An indication that *Homo erectus* was spending a good deal of time walking—perhaps running—across open habitat is evidence for heat stress adaptation. The Nariokotome skeleton exhibits the tall, slender body and long distal limbs found among modern peoples of the tropical zone. The size and shape of such a body maximizes the exposed surface area relative to overall mass and therefore allows greater heat loss. By contrast, modern peoples of the Arctic exhibit higher body mass and shorter limbs.[53] *Homo erectus* was also the first hominid to possess an external nose. This allowed moisture to be added to inhaled air and removed from exhaled air.[54] Another adaptation to heat stress may have been the loss of body hair, but this cannot be confirmed at present.

The overall increase in body size was accompanied by a reduction in sexual dimorphism. The size ratio between males and females reached the same level found among modern humans (that is, females are about 90 percent the size of males). Among living apes and other primates, reduced sexual dimorphism is associated with monogamous mating and pair-bonding. In *Homo erectus,* this development is widely thought to reflect a sexual division of labor and food sharing—related perhaps to long-range scavenging and hunting by males.[55]

The changes in anatomy and size seem to form a complex or "adaptive package" that allowed *Homo erectus* to expand its foraging range in drier and more open habitats where food resources were more broadly dispersed and variable. Humans were probably foraging over wider areas. The reduction of sexual dimorphism suggests that organizational adaptations may have been critical as well. The demands of long-range foraging would have spread hominid societies over larger areas and perhaps stretched their cohesion to its limit. Pair-bonding and food sharing may have been part of the adaptive package in these new habitats.[56]

The changes in stone tool technology that followed the appearance of *Homo erectus* may have played a less important role in the colonization of new environments. Acheulean hand axes and other large bifacial tools—which include cleavers and picks—are dated to as early as 1.65 million years in east Africa and are thus broadly correlated with the morphological changes in *Homo.* However, bifaces seem to arrive in the Near East somewhat later (roughly 1.4 million years ago) and postdate the presence of hominids by as much as 400,000 years. Hand axes are rare in most parts of East Asia, where they were clearly not critical to *Homo erectus* survival.[57]

The function of the large bifacial tools has been debated for many years. Some archaeologists have proposed that they were all-purpose tools ("Swiss Army knives" of the Lower Paleolithic). Others have suggested that they were used as both cores and tools. Experimental studies indicate that large

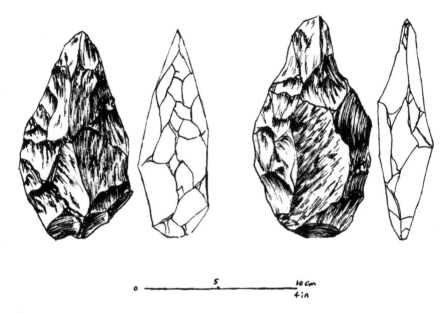

Figure 2.8. Hand axes from Olduvai Gorge in Tanzania dated to 1.5 million years ago.

bifaces are especially suited to the butchering of a mammal carcass, and there is supporting evidence from the microscopic analysis of their edges for such use.[58] It remains to be shown, however, that hand axes can perform any functions that cannot be performed by Oldowan pebble or flake tools.

The real significance of the stone bifaces may lie in their implications for human *cognitive* development. In contrast to the largely reductive technology of the earlier pebble and flake tools, hand axes, cleavers, and picks reflect a mental template or design pattern. Their production requires a sequence of manufacturing stages that transform a cobble or large flake into a shaped and symmetrical form bearing little resemblance to the original rock from which it was made.[59]

Hand axes seem to be the earliest example of humans imposing a complex structure on part of their environment. The pattern was later expanded to include a much wider array of technology and less tangible aspects of the environment, including sound and color in the form of words, art, music, and so forth.[60] It may be significant that all subsequent development of this tendency to impose complex "structure" on the natural world seems to have emerged from the descendants of the hand ax makers in Africa and Western Eurasia (that is, Neanderthals and modern humans).

Another form of technology that may have arisen at the roughly the same time is fire. Traces of controlled fire are difficult to identify in the open-air sites that dominate the archaeological record prior to 0.8 million years ago, and when identified, they are not easily distinguished from those of natural fires. Thus the lack of indisputable evidence for use of fire in this time range may simply reflect the effects of weathering and/or the limits of current techniques in archaeology. Burned bones associated with stone tools and at *Swartkrans Cave* (South Africa) and traces of possible hearths at Koobi Fora (Kenya) may indicate controlled fire as early as 1.5 million years ago.[61] If so, this technology could have played an important role in habitat expansion.

Regardless of the possible impact of new technology on the movement out of Africa after 1.8 million years ago, microwear analysis of *Homo erectus* teeth suggests a significant shift in diet. Tooth surfaces exhibit a pattern of "extreme gouging and battering" similar to that of a meat-eating and bone-crunching carnivore such as a hyena. They contrast sharply with the wear patterns observed on both living great apes and fossil australopithecine teeth.[62]

A major increase in the consumption of meat is less clear from the analysis of animal bones and artifacts at *Homo erectus* sites. Like those of *Homo habilis,* many of the archaeological sites occupied between 1.8 and 0.8 million years ago contain large mammal bones that have been broken and cut with stone tools, but indisputable evidence of hunting remains elusive. Nevertheless comparative studies of the tropical African sites reveal some subtle indications of increased meat use. At Olduvai Gorge, sites occupied by *Homo erectus* reflect greater focus on larger animals and meatier parts of the carcass.[63]

The increased consumption of meat—whether obtained primarily from scavenging or from hunting—was probably an important part of the expansion out of Africa and into higher latitudes. To begin with, the larger body and brain size placed greater energy demands on *Homo erectus.* As plant food availability declined in drier and cooler habitats, the proportion of meat in the diet must have risen. Long-range foraging and lower temperatures would have further intensified energy needs. Among recent foraging peoples, at least half of the diet of most groups above latitude 30° North is derived from hunting and/or fishing.[64] A high meat diet was essential for the subsequent colonization of the North, and—as scavenging opportunities declined at these latitudes as a function of lower animal biomass—hunting became necessary to sustain it.

CHAPTER 3

The First Europeans

The Oceanic Effect

The first human settlement of northern latitudes took place in the western half of Europe. The earliest sites—dating to 800,000 years ago or more—are known only from the Mediterranean region, but by half a million years ago, people were living as far as 50° North in Britain. Traces of their occupations are found as far east as the Danube Basin.[1]

The early phases of settlement in Europe were undoubtedly tied to the moderating effects of the North Atlantic on the westernmost part of Eurasia. Ocean currents circulating northward from the lower latitudes bring a mass of warm and moist air to Western Europe. By the time the air reaches Eastern Europe—beyond the Carpathian Mountains—it has become much cooler and drier and the "oceanic effect" has been largely dissipated. Western Europe enjoys the mildest winter temperatures and the highest biological productivity in northern Eurasia.[2]

Human geographic distribution during this interval was very similar to that of the Miocene apes between 17 and 7 million years ago (see chapter 2), and reflects in large measure the same distribution of climates. Although temperatures were lower than those of the mid-Miocene, between half a million years ago and the appearance of the Neanderthals (roughly 300,000–200,000 years ago), they were often higher than those of the present day.[3]

Indeed, environmental conditions were so favorable in Western and Southern Europe during much of this period that some have wondered why humans did not occupy them much earlier.[4] Representatives of *Homo* had colonized Eurasia as far as 41° North at least a million years before. They had coped with relatively dry and seasonally variable habitats similar to those in Western Europe, where temperatures probably approached and even exceeded the freezing point at times. This had led to speculation that a factor other than climate—possibly the extinction of several potential carnivore competitors to humans—opened the door to Europe.[5]

Adding to this speculation is the lack of evidence for adaptations to cold climates among the early occupants of Europe. Many such adaptations are present among the later Neanderthal inhabitants of the continent.[6] But the remains of the people who colonized Europe half a million years ago exhibit few if any differences from those of earlier humans that would have

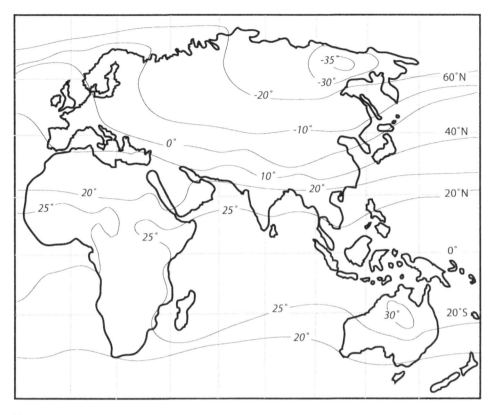

Figure 3.1. Map of winter temperatures (°C) in northern Eurasia today, illustrating both latitudinal and longitudinal climate gradients.

helped them cope with cold environments. The first movement into the higher middle latitudes—ultimately as far as 52° North—may have been no more than an opportunistic extension of the human geographic range under favorable circumstances.

Despite the lack of obvious cold adaptations in their record, there is reason to believe that the earliest Europeans may have developed some special means of living in higher-latitude environments that set them apart from their southern contemporaries. However ameliorated by local conditions and warm-climate oscillations, Western Europe must have presented some challenges to its first human inhabitants. Moreover, there is at least some evidence for occupation during periods when climates were actually colder than the present day.[7] Our inability to detect cold adaptations may be due largely to the fragmentary character of the record. Human skeletal remains are particularly scarce from this part of the world prior to 300,000–200,000 years ago, and the early Europeans may have evolved some anatomical responses to low temperature that have yet to be found. Other adaptations—such as an increased consumption of meat or expanded use of mammal hides—also might be difficult to recognize among the limited debris of the archaeological sites.

Initial Occupation

The human colonization of Europe took place in at least two stages. After half a million years ago, settlement is widespread and well documented in both the northern and the southern regions of Western Europe. Sites in most regions yield hand axes and the skeletal remains of people who appear to have been ancestral to the Neanderthals. But prior to 500,000 years ago, the pattern of settlement is different. Firmly dated sites are extremely rare and currently confined to southern Europe. Although scarce, human skeletal remains must be assigned to other hominid taxa. Hand axes and other bifacial tools are largely—if not wholly—absent.[8]

The earlier phase of occupation, which began at least 800,000 years ago, appears to represent one or more colonization events by relatively small numbers of humans. Both their skeletal morphology and their tools suggest that they may have had little connection with the people who colonized Europe after 500,000 years ago, and they might have failed to establish long-term settlement. The lack of known sites in northern Europe suggests that the initial occupants may have been unable to cope with environments above 41°–42° North (that is, above latitudes already settled by *Homo erectus* in Asia).

Documenting the early phase of European colonization is difficult because of its limited visibility in the archaeological record. The density of the early European population was probably low, and occupation sites may have been small. Few of the sites are likely to have been preserved, and even these may be particularly difficult to find. Most caves and rock shelters—which protect archaeological remains and are easy to identify as potential sites—erode away in a few hundred thousand years. The majority of the early European sites are buried in sediments deposited by streams, lakes, or springs.[9]

The lack of hand axes presents a special problem for the European sites that antedate half a million years. Such tools are unmistakable products of the human hand, and their presence in later deposits—even in isolated settings—is firm evidence of human occupation. But prior to 500,000 years ago, Europeans were making simple pebble and flake tools that differed little from the original Oldowan industry (see chapter 2). These artifacts are often difficult to distinguish from naturally fractured rock, and they are frequently recovered from geologic contexts (such as high-energy

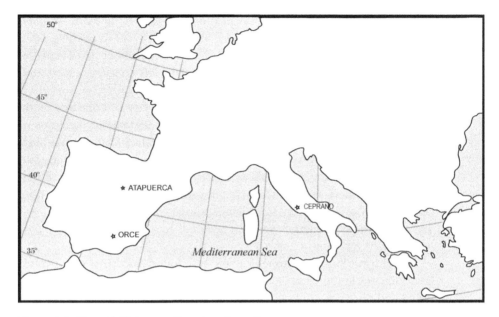

Figure 3.2. Map of initial occupation of southern Europe.

stream deposits) likely to contain naturally chipped and broken cobbles and pebbles.[10]

As a consequence, most of the reported European sites dating to more than 500,000 years ago are highly problematic. For example, artifacts have been reported from several localities dating to as much as 1.8–2.0 million years ago in the mountainous Massif Central region of France. The purported tools include stone choppers, polyhedrons, and modified flakes. Comparative analysis, however, reveals that similar pieces are formed by volcanic processes, which were once common in the Massif Central. Further east, the Czech site of *Prezletice* near Prague has yielded several hundred chipped fragments of lydite and mammal remains in ancient lake deposits dating to roughly 700,000 years ago. But the fragments lack the characteristics of humanly flaked stone, and many archaeologists believe they represent naturally fractured rocks. Similar examples are known from other parts of Europe.[11]

Sierra de Atapuerca

The oldest widely accepted evidence for the occupation of Europe is found at the site of Atapuerca in northern Spain (latitude 42° North). Atapuerca is a rare example of a cave that is more than half a million years old. In fact, the *Sierra de Atapuerca* is a mountain composed largely of limestone that is honeycombed with tunnels and cavities (or *dolinas*). One of them—named the "Gran Dolina"—is filled with more than 50 ft (16 m) of pebbles, sand, silt, and clay deposited primarily by running water. In a layer that is dated to about 800,000 years ago, Eudald Carbonell and his colleagues recovered both stone artifacts and human skeletal remains.[12]

The human remains comprise eighty-five fragments of bone and teeth, representing parts of both the cranial and the postcranial skeletons of at least six individuals. They exhibit some anatomical features that align them with *Homo erectus* of Africa and Asia, and even with modern humans *(Homo sapiens)*, but set them apart from the later European hominids. Among these features is a pronounced depression between the nose and cheek bone *(canine fossa)* that is absent in the Neanderthals. Accordingly, these remains have been assigned to a separate species *(Homo antecessor)*.[13]

Roughly two hundred stone artifacts were found associated with the human remains in the Gran Dolina. They are primitive-looking cores, pebble tools, and retouched flakes of limestone, sandstone, and flint that are generally similar to the Oldowan industry. Smaller assemblages of similar pieces were also recovered from underlying layers that have not yielded any human bones.[14] Had the Atapuerca artifacts from all these levels been found without human remains, they might have been widely dismissed as naturally fractured rock.

Atapuerca is a mysterious site, and it is not clear precisely how the human bones and artifacts came to be deposited in the cave. A number of large mammal remains were found in the same levels, including those of bear, hyena, elephant, horse, deer, and others. Bears are especially common in one of the lower layers, where whole skeletons were found intact and probably represent animals that died during hibernation. Some mammal bones were gnawed by carnivores, and may be the remains of prey brought to the cave by nonhuman predators; others exhibit traces of tool damage. In fact, the human bones also reveal cut marks from stone tools and seem to be early evidence of violence and cannibalism.[15]

Southern Spain may contain other isolated traces of early hominid occupation. Crude pebble tools and flakes are known from lakeshore deposits that may be as much as a million years old in the Orce Basin near Granada.[16] Aside from Atapuerca, however, the most important evidence is found in central Italy at *Ceprano*. This site, which is located roughly 55 miles south of Rome, lies at almost the same latitude—slightly below 42° North—as Atapuerca. In 1994 clay deposits exposed by a road-building project near Ceprano yielded a human braincase (or calvarium). The skull fragment, which is thought to be older than 700,000 years, exhibits many similarities with Asian *Homo erectus*.[17]

Heidelberg Man

Between 800,000 and 525,000 years ago, the northern hemisphere experienced three periods of warm climate (interglacials) interspersed with colder episodes, and it appears likely that the sporadic colonization events of this interval occurred during the warm periods. After the beginning of the interglacial that commenced roughly 525,000 years ago, the pattern of European settlement changed radically. From that point onward—until the gradual emergence of the Neanderthals after 300,000 years ago—the western and southern parts of the continent were inhabited on a more or less permanent basis by a distinct group of people, who are widely classified as *Homo heidelbergensis*.[18]

The sites occupied by *Homo heidelbergensis* are relatively common though by no means abundant in the warmer regions of Europe. With the exception of the southernmost edge, the colder and drier portions of the continent that lie north and east of the Carpathian Mountains remained uninhabited (sites are scarce in regions where the current winter temperature mean falls below the freezing point). The majority of sites were occupied during the warm interglacial periods, but there is also some evidence of habitation during the colder intervals. The human population may have declined in numbers and distribution at these times—in contrast to the later Neanderthals, who thrived in cold climates.[19]

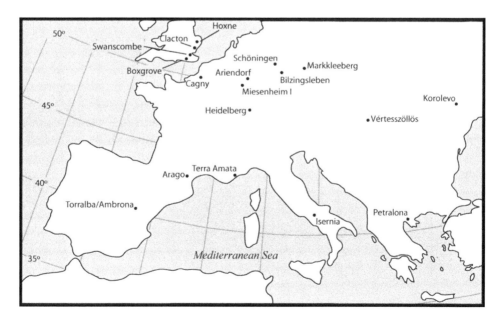

Figure 3.3. Map of sites occupied by *Homo heidelbergensis*.

Box 1. Climate Change in Europe: 800,000–300,000 Years Ago

The initial settlement of higher latitudes—the occupation of Western Europe during the earlier Middle Pleistocene—took place under climate conditions that were generally warmer than those of the present. Between 800,000 and 300,000 years ago, a series of major warm intervals (interglacials) took place. These were interspersed, however, with episodes of varying cold (glacials).

The general framework for climate change during this period is provided by the oxygen-isotope record of deep-sea cores. These cores measure past fluctuations in the ratio of ^{18}O to ^{16}O in marine fossils (typically foraminifera). As glaciers expanded during cold periods, seawater became enriched with higher concentrations of the heavier isotope (^{18}O). When ice volume shrank during the interglacials, the oceans were flooded with glacier meltwater containing higher percentages of the lighter isotope. The deep-sea record is dated in part by paleomagnetism and correlates with some deposits on land, including the wind-blown silt (loess) stratigraphy of northern Europe.[a]

According to the calibrated oxygen-isotope record, there were at least six major interglacials between 800,000 and 300,000 years ago. The warm intervals are designated with odd numbers (for example, oxygen-isotope stage 19 or OIS 19), and especially warm interglacials are dated to roughly 425,000–360,000 years ago (OIS 11) and 340,000–300,000 years ago (OIS 9). During these intervals, temperatures were at least a few degrees warmer than those of today, and many of the areas of Western Europe occupied by humans supported a temperate oak woodland. Mammals at the archaeological site of Hoxne in England (tentatively dated to OIS 9) included red deer, horse, and elephant.[b]

Fitting human fossils and sites into the climate record for Europe prior to 300,000 years ago is a challenge, and many of them cannot be assigned to a specific oxygen-isotope stage. Nevertheless, at least some sites appear to date to even-numbered glacial intervals (such as OIS 12). Many archaeologists believe that the range of climate tolerance among the early Europeans was relatively broad and that—at times—people were living in much cooler environments.[c]

---

a. Martin J. Aitken, "Chronometric Techniques for the Middle Pleistocene," in *The Earliest Occupation of Europe,* ed. W. Roebroeks and T. van Kolfschoten, pp. 269–277 (Leiden: University of Leiden, 1995); A. A. Velichko, "Loess-Paleosol Formation on the Russian Plain," *Quaternary International* 7/8 (1990): 103–114.

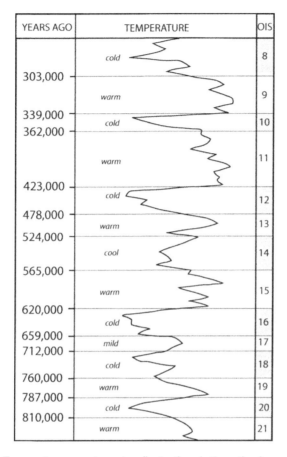

Figure B1. Temperature curve based on fluctuations in the ratio of oxygen isotopes from a deep-sea sediment core.

b. N. J. Shackleton and N. D. Opdyke, "Oxygen Isotope and Paleomagnetic Stratigraphy of Equatorial Pacific Core V28-238: Temperatures and Ice Volumes on a 10^3 and 10^6 Year Scale," *Quaternary Research* 3 (1973): 39–55; Ronald Singer, Bruce G. Gladfelter, and John J. Wymer, *The Lower Paleolithic Site at Hoxne, England* (Chicago: University of Chicago Press, 1993).

c. Clive Gamble, "The Earliest Occupation of Europe: The Environmental Background," in *The Earliest Occupation of Europe,* ed. Roebroeks and van Kolfschoten, p. 281.

The people who appeared in Europe half a million years ago had close ties to Africa, and this is clear from both their artifacts and their skeletal remains. A new wave of migration out of Africa seems to be represented at *Gesher Benot Ya'aqov* in Israel about 750,000 years ago. The occupants of this site were making hand axes very similar to those found in African sites of comparable age (and less so than those of earlier Near Eastern sites).[20] A human cranium found many years ago in South Africa *(Broken Hill)*—and estimated to be roughly 400,000 years old—bears a remarkable resemblance to a cranium discovered in Greece *(Petralona)*. Both have been classified as *Homo heidelbergensis*.[21] By 500,000 years ago, these people were making hand axes at *Boxgrove* in southern England above latitude 50° North.[22]

The taxon *Homo heidelbergensis* was proposed by Otto Schoetensack in 1908 on the basis of a isolated jaw bone recovered from a sand quarry near Heidelberg during the previous year. Its stratigraphic position and associated interglacial fauna indicate that it is about the same age as the Boxgrove site. The Heidelberg mandible is massive and lacks a chin, but the teeth are comparatively small.[23] It has long been viewed as intermediate between *Homo erectus* and modern humans, and it now seems to represent a species that was present in both Europe and Africa half a million years ago and became the common ancestor of Neanderthals and modern humans.[24]

Although almost a full century has passed since the original Heidelberg find, the number of bones and teeth that may be assigned to this taxon remains quite limited. The nearly complete cranium from Petralona is perhaps the most important specimen. Like the Heidelberg jaw, it retains many primitive features, including a heavy brow ridge and a relatively low and flat cranial vault, but it exhibits a much larger brain volume—about 1,220 cc—than *Homo erectus*. A partial but rather distorted cranium, along with two mandible fragments and several other bones and teeth, has been recovered from *Arago* in southern France. This is one of the few known cave occupations from this time period and it probably dates to about 400,000 years ago (thus younger than Boxgrove and the Heidelberg jaw). Nevertheless, the Arago remains are similar in many ways to the older fossils and indicate brain size comparable to the Petralona skull.[25]

As noted in the preceding chapter, modern peoples of the Arctic tend to possess greater body weight and shortened limbs. This reflects a general pattern observed by nineteenth-century naturalists among northern representatives of warm-blooded animals. By reducing the exposed surface area of the body, they conserve heat and protect against cold injury. Other adaptations to low temperature include insulation in the form of subcutaneous fat and thick coats of fur or feathers.[26] Although later inhabitants of Europe—the Neanderthals—exhibit anatomical adaptations to cold in the

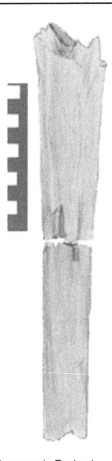

Figure 3.4. Human tibia from Boxgrove in England.

form of large body mass and shortened limbs, evidence of such adaptations is lacking for *Homo heidelbergensis*.

Most of the sample of skeletal parts is confined to skull fragments and teeth, which unfortunately provide little information about climate adaptation. Although increased head size reduces heat loss,[27] the expansion of brain volume observed in *Homo heidelbergensis* is present among the African specimens and is probably unrelated to latitude. An exception is a lower leg bone (tibia) recovered in 1993 from Boxgrove. The specimen had to be reconstructed from fragments and is missing both its proximal and its distal ends, but provides a basis for estimating the length of the original lower leg bone, as well as the size and weight of the individual.

The Boxgrove tibia is currently thought to have been 15–16 inches in total length (375–400 mm) and to have belonged to an adult male roughly 6 feet in height (1.77–1.82 m). This suggests a relatively tall individual with long limbs—more typical of modern people from southern latitudes and comparable even to the early *Homo erectus* skeleton from Nariokotome in Kenya (see chapter 2).[28]

At the same time, the large circumference of the Boxgrove tibia shaft suggests a total body weight of more than 200 lbs (80 kg), which is substantially heavier than the mean for adult males among tropical peoples.[29] The Boxgrove specimen thus delivers a mixed message. Adaptation to cold climate in *Homo heidelbergensis* is not evident so far in limb dimensions, but might be reflected in increased body mass. However, a much larger sample of postcranial remains is required to assess properly the role of morphology in the early settlement of Europe. Furthermore, it should be kept in mind that morphological adaptations to northern climates could have included soft parts—such as a thicker coat of body hair—that are not preserved in the fossil record.

Diet and Ecology

In a classic 1992 paper, Alan Turner suggested that the critical factor in the widespread occupation of Europe roughly half a million years ago was the extinction of competitor species. He argued that the disappearance of several large carnivores about 500,000 years ago, including the giant cheetah *(Acinonyx pardinensis)* and the lesser scimitar cat *(Homotherium latidens)*, would have opened up an ecological niche in Europe for hominids by increasing the quantity of available mammal carcasses for scavenging.[30] This thesis implies that novel adaptations to higher latitudes may not have played a role in the colonization of Western Europe.

As already noted, the small sample of *Homo heidelbergensis* skeletal remains provides little evidence for anatomical adaptations to cooler climates. Evidence for changes in diet and ecology is also rather limited and ambiguous. The best source of information on diet lies in human bone chemistry and the values yielded for specific stable isotopes that reflect the percentages of consumed plant and animal food. Stable isotope values for the Neanderthals and modern humans in Europe reveal heavy consumption of animal foods (see chapters 4–5). No such data are yet available, however, for *Homo heidelbergensis,* and knowledge of their diet is almost entirely based on the analysis of animal and plant debris recovered from archaeological sites.

Recently published results from Boxgrove suggest that the early Europeans were probably consuming significant quantities of meat and marrow obtained from large mammals. Boxgrove is especially important, because it was occupied primarily during a warm interglacial period (when plant foods were most abundant) at the time that humans first appeared above 42° North. Careful analysis with a scanning electron microscope revealed many traces of stone-tool cut marks on bones of horse, red deer, rhinoceros, and other large mammals from this site. Impact fractures were also observed on a number of bones—presumably smashed open to extract the marrow. The locations of the cut marks indicate that the Boxgrove people had early access to complete carcasses. In some cases, the tool cuts were overlaid with tooth marks, suggesting that carnivores were chewing the bones only after humans had discarded them.[31]

Traces of stone tool cut marks and impact fractures have been found on mammal bones from other European sites dating to between 500,000 and 300,000 years ago, including *Schöningen* (Germany), *Cagny l'Epinette* (France), and *Isernia La Pineta* (Italy).[32] But there is little or no evidence for the dismemberment and butchering of carcasses at some localities, where it appears that most of the large mammal remains may have accumulated by natural processes. Examples include *Miesenheim I* in the Rhine Valley (Germany) and *Treugol'naya Cave* in the northern Caucasus Mountains (southern Russia).[33] Furthermore, microscopic analysis of wear on the teeth of the Heidelberg jaw indicates that abrasive plant foods were still a major part of the diet.[34]

The overall pattern suggests that while people could sometimes obtain large amounts of meat in the form of an entire carcass, the percentage of animal foods in their diet might have been much lower than that of Neanderthals and modern humans in northern environments. As described in chapter 2, both the eating of meat and the cutting and breaking of large mammal bones is documented in African hominid sites as early as 2 million years ago. The question is therefore—did the people who colonized Europe half a million years ago significantly increase their consumption of meat as a means of coping with cooler climates and reduced plant abundance?

Although this question cannot be answered decisively with current data and available methods, for all of the reasons mentioned in the preceding two chapters, it seems almost certain that humans had increased their meat intake since *Homo habilis* in Africa. The decline in available plant foods must have been considerable—on the basis of comparisons between modern tropical and northern temperate woodlands—especially during the winter months in Europe.

At *Hoxne* in England, climate conditions during the period of occupation (roughly 300,000 years ago) were estimated using the remains of beetles, which are sensitive indicators of temperature. While mean July temperatures appear to have ranged from 59° to 67° F (15°–19° C), the January mean is estimated at between 14° and 44° F (−10°–6° C).³⁵ Clearly, this is significantly cooler—and with far greater seasonal extremes—than the tropical zone, but probably no more so than regions of midlatitude Eurasia occupied earlier. Humans may have adjusted to the change with increased meat consumption as they expanded into these drier and more seasonal environments between 1.8 and 0.8 million years ago (see chapter 2). Perhaps, as Turner implied, the move into Europe required no fundamental changes in diet.

A closely related question is the importance of hunting. How much of the meat consumed by *Homo heidelbergensis* was hunted rather than scavenged? In the early 1980s archaeologists became skeptical of the assumption that early humans had always been big-game hunters. More rigorous methods were applied to the study of large mammal remains from Lower and Middle Paleolithic sites in an effort to sort out their history. For example, elephant bones recovered from *Torralba/Ambrona* in Spain during 1961–1963 were originally thought to represent the food debris of organized kills.³⁶ But later analysis of the bones revealed a complex history of stream deposition, carnivore activity, and perhaps only limited human involvement.³⁷ Here—as at other Lower Paleolithic sites studied after 1980—it was particularly difficult to distinguish between hunting and scavenging.

Nevertheless, it seems increasingly likely that hunting accounted for a substantial portion of the meat in the *Homo heidelbergensis* diet. One reason for this is that while large mammal carcasses accumulate in significant quantities on tropical woodland and savanna landscapes, their numbers decline per unit area as plant productivity falls off.³⁸ Scavenging opportunities for the early Europeans would have been considerably less than those for *Homo habilis*. Furthermore, the Neanderthals—who were the immediate successors of *Homo heidelbergensis*—were heavy hunters of big game and it is improbable that they initiated the practice of hunting.

If the cut and broken mammal bones at sites like Boxgrove and Schöningen were the remains of hunted prey, it also is likely that the early Europeans were practicing a strategy of "central-place foraging." Here again, however, it is difficult to confirm this with current data and methods. Most sites in this time range are found in settings where large mammals and/or their remains could have been accumulated by natural processes. At sites like *Vértesszöllös* (Hungary), where bones and artifacts are concentrated around an ancient spring, or Hoxne, where occupation debris was deposited along stream and lake margins, it is not clear that people brought food items back to a home base.³⁹ This might be partly another consequence of the scarcity of caves in this period: there is ample evidence of central-place foraging among the many caves and rock shelters occupied by Neanderthals and modern humans in Western Europe.⁴⁰

Technology

One of the principal means by which modern humans adapt to cold environments is technology. There is a strong correlation between latitude and the diversity and complexity of tools and facilities among recent foraging peoples. Complicated technology for clothing and shelter alone were obviously critical for survival in subarctic and arctic habitats. But while there is some evidence for increased technological sophistication among the Neanderthals, the earlier inhabitants of Europe may have lived without any novel devices or techniques.

The technology of *Homo heidelbergensis* in Europe—like its morphology and ecology—is hard to differentiate from that of its contemporaries in lower latitudes. Before the emergence of the Neanderthals roughly 300,000 years ago, a unique phase of human prehistory ensued during which both the artifacts and the people of tropical Africa seem indistinguishable from those of temperate Europe. Both populations were producing essentially the same array of stone bifaces and flake tools, along with some items of wood and (probably) hide. As in the case of morphology and ecology, however, some differences might have existed that have little or no visibility in the archaeological record.

The hand axes of *Homo heidelbergensis* were the first artifacts of the Old Stone Age to be recognized as such. In 1797 John Frere reported their discovery at Hoxne to the Society of Antiquaries in Britain, and in the nineteenth century they became instrumental in establishing the antiquity of humankind.[41] The hand axes made by the early Europeans reflected the refinement of the later Acheulean industry. Many specimens were shaped with beautiful precision and exhibit symmetry in both plan and profile (that is, in three dimensions). As noted in the preceding chapter, their primary significance may lie in their form—an abstract structure imposed on the natural world—rather than in their function(s). Other large bifacial tools included cleavers and picks.[42]

Some of the European sites have provided evidence that hand axes and other bifaces were used to butcher large mammal carcasses. At Boxgrove, where so many of the large mammal bones exhibit tool damage, hand axes completely dominate the artifact assemblage. Many of the cut marks on the bones seem to have been made with the edge of a bifacial tool.[43] At Hoxne, where cut marks were also observed on large mammal bones, microscopic analysis of hand axes revealed traces of meat-cutting wear on their edges.[44]

In addition to the large bifacial tools, *Homo heidelbergensis* produced simple stone flakes that were used—according to microscopic analysis of their sharp edges—to cut meat, hide, and wood. Some flakes were retouched into simple scraping and cutting tools, and these were also used on

Figure 3.5. Wooden spear (or probe?) from Schöningen in Germany.

a variety of materials, including hide, wood, nonwoody plants, and bone. Compared with the finely made bifaces, the flake tools are less standardized, and the retouching of edges was often—it appears—designed to make them easier to hold in the hand.[45]

Although there are no shaped tools of bone, antler, or ivory tusk in these sites, several wooden implements have been recovered. The German site of Schöningen yielded three spears or sharpened poles of spruce (5.9–7.5 ft [1.8–2.3 m] in length) and a shorter stick sharpened at both ends. Many years ago, the broken shaft of a possible spear, with a pointed end, was found at *Clacton-on-Sea* in England. This specimen was fashioned from yew and exhibits traces of microscopic working along the point.[46] Wood is rarely preserved in deposits of such age (400,000–300,000 years old), and the fact that a specimen was found, combined with the microscopic evidence of woodworking on flakes and flake tools, suggests that wooden tools—at least of simple design—were common.

Evidence for the scraping of hides from microwear studies of tools from Hoxne and Clacton-on-Sea may indicate the presence of protective clothing. Because evidence for tailored fur clothing in the form of eyed needles and other items associated with its production among tribal peoples is confined to modern humans (see chapter 5), the clothing of *Homo heidelbergensis* must have been simple—probably limited to wraps, ponchos, and blankets. Such clothing would have provided little protection against very low temperatures (except perhaps while sleeping).

Construction of simple shelters or huts has been reported from several European sites in this time range (most notably *Terra Amata* in southern France), but most archaeologists are highly skeptical of this evidence.[47] Overall, technology related to clothing and shelter seems to have been very limited.

Undoubtedly the use of controlled fire remains the greatest technological mystery—and one with special significance for cold-climate adaptation—among the early Europeans. As recounted in the preceding chapter, there is some evidence that fire was used as early as 1.5 million years ago in Africa. Traces of fire use in archaeological sites more than roughly 250,000 years ago, however, is rare and almost always subject to dispute. Remains of structured hearths are completely absent and the evidence for controlled fire is limited to burned bones (for example, Vértesszöllös, Hungary), burned-wood fragments (*Bilzingsleben*, Germany), and isolated charcoal fragments (*Swanscombe*, England). In each case, alternative explanations have been proposed, and the issue is unresolved. Of special significance may be the absence of hearths or other traces of fire at Arago in the French Pyrenees, where they are more likely to have been protected and preserved in the cave.[48]

A few years ago, the British prehistorian Clive Gamble suggested that the early Europeans might have employed some of their technology to create a new ecological niche for themselves. Noting that the winter months at higher latitudes would have placed a particular strain on human foraging, Gamble proposed that humans could have used wooden probes (such as the implements subsequently found at Schöningen) to locate the carcasses of large mammals buried under the snow. These would have accumulated as a result of natural deaths during the late fall and early winter period, becoming inaccessible to carnivores as they froze and disappeared beneath the falling snow. Once the carcasses were located, humans could have thawed and dismembered them with the use of fire.[49] On the other hand, if *Homo heidelbergensis* had become a relatively proficient hunter, scavenging might have been unnecessary to overcome the problems of winter food shortage.

Over the Line: Central Europe

In 1948 Hallam Movius delineated a boundary across Eurasia separating Paleolithic hand ax industries from those without hand axes. The "Movius Line," as it became known, ran from the Caucasus Mountains across south Central Asia and northeast India. Sites containing hand axes and other large bifacial tools were restricted to regions west of the line, while chopper and flake-tool industries were to be found east of the line.[50] Ever since then, archaeologists have been trying to understand the meaning of the Movius Line.

The dating of the earliest human remains and archaeological sites in Eurasia to as much as 1.7 million years ago has altered the picture in recent years. It now seems that the oldest industries outside Africa—including perhaps those east of the Movius Line—antedate the hand ax, which first appears in Africa about 1.65 million years ago. In fact, hand axes do not show up in Eurasia until about 1.4 million years ago. Furthermore, recent dates on archaic human fossils in the Far East now suggest that there may have been relatively little evolutionary change there until the arrival of modern humans.[51] Thus the simplest explanation of the Movius Line is that it

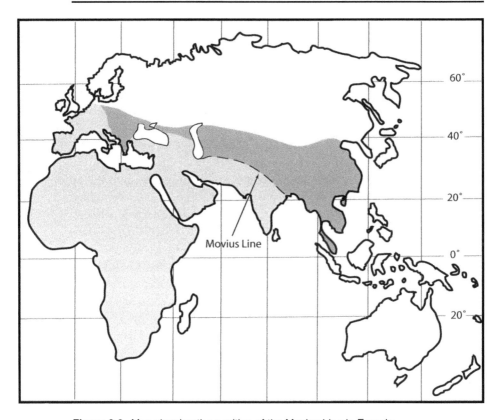

Figure 3.6. Map showing the position of the Movius Line in Eurasia.

reflects the early movement of *Homo* into eastern Eurasia with pre–hand ax technology that remained essentially unchanged until the spread of modern humans.

It also appears that the Movius Line extends north of 42° into Central Europe. Pre-Neanderthal sites east of Switzerland—or roughly longitude 12° East—contain chopper and flake-tool assemblages like those of the East Asian industries. The sites are much less common than those in Western Europe, perhaps because they represent the margins of the human geographic range at this time. They include Bilzingsleben and Schöningen in eastern Germany and Vértesszöllös in Hungary, as well as the site of *Korolevo*, which lies at the easternmost edge of the Danubian Basin.[52]

The human skeletal remains associated with these sites reinforce the impression that they form a group distinct from that of the West European sites. Fragments of the back of the skull (occipital bone) from both Bilzingsleben and Vértesszöllös exhibit very primitive features and are said to resemble *Homo erectus*. This has encouraged speculation that a separate human species—with ties to the East Asian chopper-tool makers—might have inhabited Central Europe.⁵³ Unfortunately, the sample of remains is too small to resolve this issue at present.

An alternative explanation for the absence of hand axes in the Central European sites is that these sites simply reflect functional differences with the hand ax sites—or the scarcity of suitable local materials for hand ax production. In fact, some of the West European sites also lack large bifacial tools, including Clacton-on-Sea and Isernia La Pineta. Many archaeologists suspect that the absence of hand axes in these sites has nothing to do with biological or geographic differences among human populations. However, there are no obvious functional differences between the two sets of sites, and the pattern remains a puzzling one.⁵⁴

More important here is the fact that climates become increasingly continental—with significantly lower winter temperatures—in Central Europe. Charles McBurney, who was one of the first to call attention to the Central European industries without hand axes, attributed their character to harsher local climates.⁵⁵ This raises the intriguing possibility that the elusive evidence for cold adaptation in *Homo heidelbergensis* might be found in the contrast between the regions.

It is difficult to account, however, for the absence of hand axes in terms of climatic differences. In fact, given the apparent link between bifaces and butchery on the one hand and the relationship between meat consumption and climate on the other, one might expect *more* hand axes in Central Europe. Nevertheless these sites do yield much evidence for controlled fire and tool cut marks on mammal bones—both of which could reflect responses to cooler environments—but there are too many uncertainties about these data to confirm that a pattern is present.⁵⁶ Like the chopper and flake-tool assemblages of Western Europe, the significance of the Central European sites is unclear.

Before the Neanderthals

The earliest stage of the settlement of latitudes above 42° remains the most poorly known and understood in the prehistory of the North. The people who occupied Western Europe prior to the emergence of the Neanderthals are still an enigma in many respects. This is almost certainly due to the great antiquity of their sites and the consequent loss of materials over time. In contrast to what is known of the later occupants of Europe, we still lack critical information regarding both their physical appearance and their way of life.

Evidence for anatomical or behavioral adaptation to northern environments among the early Europeans is virtually nonexistent. Instead, a period unique in prehistory would seem to have ensued after 525,000 years ago, when the human inhabitants of Africa and Europe were essentially indistinguishable in both respects. This period came to an end roughly 300,000 years ago with the emergence of northern and southern forms of *Homo*.[57]

Alan Turner's thesis that carnivore extinctions opened an ecological niche for *Homo heidelbergensis* in Europe approximately 500,000 years ago implied that humans might have been "pre-adapted" for the warmest parts of the continent.[58] Although the archaeological record remains murky, both hunting and the use of controlled fire may have been important components of the earlier spread across Eurasia. The use of hides for blankets and simple clothing also may have been part of life in southern Eurasia prior to 500,000 years ago. Perhaps these adaptations permitted humans to extend their range into Western Europe at that time.

But an equally plausible possibility is that *Homo heidelbergensis* developed one or more novel adaptations to higher latitudes that have yet to be teased out of the fragmentary record. These novel adaptations could have included:

1. Increased body size and thicker body hair
2. Physiological responses to low temperatures
3. Increased consumption of meat (related to intensified hunting)
4. Expanded wood and hide technology (including hunting weaponry and clothing)
5. Expanded use of controlled fire

Such adaptations seem more likely—if not inevitable—if *Homo heidelbergensis* managed to sustain itself in Europe during the two major cold intervals that occurred between 500,000 and 300,000 years ago. Mean winter temperatures would have fallen significantly below those of today during these glacial episodes.[59]

Although the majority of sites date to phases when climates were as warm as or warmer than the present, there are at least several significant exceptions. Traces of occupation during the earlier glacial interval are reported from the upper levels at Boxgrove in England and from the lower levels at *Cagny-Cimetière* in northern France. Occupations dating to later glacial periods are thought to be present at Arago, *Ariendorf* on the Rhine, and also at *Markkleeberg* in eastern Germany.[60] If additional evidence for glacial-phase habitation accumulates, the case for cold adaptations in *Homo heidelbergensis* will become stronger.

CHAPTER 4

Cold Weather People

The Neanderthals evolved gradually in Western Europe from *Homo heidelbergensis*. By 300,000 years ago, many of their characteristic features are visible in the European fossil record. They subsequently expanded the human geographic range into the colder and drier regions of Eastern Europe and even further east into parts of Siberia. In contrast to their predecessors, they occupied numerous sites during full glacial episodes in at least some of these regions.

The Neanderthals were a specialized northern form of *Homo*. Many of their anatomical features seem to have evolved as adaptations to a cold climate, and their diet was heavy in protein and fat like that of modern peoples at high latitudes. Both their morphology and diet must have been critical to their expansion into cold environments never previously inhabited by humans. They made some advances in technology as well, although in this respect they seem to have been far behind the modern humans who later occupied these settings.[1]

The Neanderthals were first discovered by a group of quarrymen in August 1856 in a small cave in the Neander Valley (or *thal* in the old German spelling) near Düsseldorf. The remains, which included the top of a skull and several limb bones, were handed over to a local schoolteacher, who brought them to wider attention. Although Darwin was unaware of the Neanderthal discovery when he published *The Origin of Species* three years later, Thomas Henry Huxley described the skull cap in a book of essays published in 1863. During the same year, the Irish anatomist William King proposed that the remains be classified as *Homo neanderthalensis*—an extinct species of human.[2]

Between 1886 and 1914 many new Neanderthal discoveries were made in Germany, France, and Belgium, and some remains were also found in Croatia. Following World War I, fresh discoveries were made in Italy and Ukraine, and also in the Near East and Central Asia. Today several hundred individuals are represented, and the Neanderthals have become the most thoroughly known of archaic humans.[3]

Perhaps in part because of the wealth of finds, many controversies surround their interpretation. The principal focus of debate concerns the evolutionary relationship between the Neanderthals and modern humans. Some paleoanthropologists argue that living Europeans are at least partly descended from the Neanderthals as a result of significant interbreeding with modern humans. Others believe that modern humans—dispersing out of Africa and into Europe 50,000–40,000 years ago—effectively replaced the local Neanderthal population with little or no genic exchange.[4] Since the late 1980s the debate has been enlivened by a growing amount of biomolecular data, including mitochondrial DNA studies of modern humans and analysis of fossil DNA samples from several Neanderthals.

Other controversies about the Neanderthals concern the degree to which their thought and behavior were similar to those of modern humans. Many view the Neanderthals as uniformly primitive—despite their large brains—lacking the capacity for language, abstraction, and planning. Although the Neanderthals often buried their dead, some scholars have suggested that this was simply a disposal of corpses without symbolic or ritual meaning. But others believe that the Neanderthals became increasingly similar to modern humans—especially the first modern humans in Europe—and that the transition was a smooth one. These sharply contrasting interpretations of Neanderthal behavior have made their way into popular novels and films about these mysterious people.[5]

One of the reasons why the Neanderthals are so difficult to understand is that they are not ancestral to modern humans, but rather the product of a parallel and separate line of evolutionary development. In some respects, they may be considered an alternative form of modern human. Just as they evolved certain unique anatomical features, the Neanderthals probably developed some peculiar patterns of behavior that never appeared in modern humans and are unknown among earlier hominids. Their burial of the dead—without convincing evidence of ritual—could be an example of this. They may have developed some unique forms of communication and organization as well. Such possibilities present a formidable challenge to archaeologists.

Neanderthal Origins

The process of speciation—the means by which one species becomes another—is often thought to entail rapid evolutionary change as an organism abandons one integrated complex of traits and converges on a new one. For the last few decades, biologists have been debating this model ("punctuated equilibria") against the "phyletic gradualism" envisioned by Darwin.[6] But while it is possible—given the inevitable gaps in the fossil record—that minor bursts of rapid change occurred during the process of their evolution, the Neanderthals seem to be a classic case of gradualism that took place over a period of several hundred thousand years.

In July 1997 Svante Pääbo of the University of Munich and several colleagues published the results of a remarkable study in which they had managed to extract and analyze DNA from one of the bones found in the Neander Valley in 1856. Since 1997, more Neanderthal DNA has been analyzed from specimens recovered in Croatia and southern Russia. In all cases, the fossil DNA revealed significant genetic distance from modern humans. The results of these studies were discussed primarily in terms of their implications for Neanderthal extinction, but they also shed light on Neanderthal origins. On the basis of the measures of genetic distance, Pääbo and his colleagues estimated that the Neanderthal and modern human lineages diverged between 690,000 and 550,000 years ago.[7] This accords well with the human fossil record, supporting the notion that both lineages are descended from *Homo heidelbergensis*.

Characteristic Neanderthal traits show up in the European fossil record as early as 500,000 years ago. The small size of the cheek teeth—relative to the front teeth—in the Heidelberg jaw may be one of the earliest examples. The shape of the face on the slightly younger Arago specimen (roughly 400,000 years ago) is thought to be significant. These traits accumulated during the course of the next quarter of a million years, and eventually defined the Neanderthals as a distinct human species. The dividing line between *Homo heidelbergensis* and *Homo neanderthalensis* is a blurred and arbitrary one, and some anthropologists extend the latter classification back to the earlier fossils.[8]

Of special importance is the emergence of traits that evolved as part of the Neanderthal adaptation to a cold climate. These traits seem to have appeared during the later phases of the transition. Most of the Neanderthal features recognized in the earlier phases seem to be largely the product of isolation from the African population and genetic drift (although—as noted in the preceding chapter—it is difficult to confirm this with such a small sample of postcranial remains). But after 300,000 years ago, climates were cooler in Europe and subsequent change was probably driven by selection for cold-climate features.[9]

An exceptionally large human fossil assemblage has recently been recovered from Atapuerca in northern Spain (see chapter 3). Deposits in the *Sima de los Huesos* ("Pit of the Bones") are younger than those that produced the earliest European fossils—probably about 320,000 years old. They have yielded more than two thousand remains, including jaws, crania, ribs, limb bones, and other skeletal parts. The Sima de los Huesos assemblage exhibits many typical and unique Neanderthal features, such as a large gap between the last molar and the ascending ramus of the jaw *(retromolar space)* and a depressed area on the back of the cranium *(suprainiac fossa)*. But the remains

Box 2. Climates of the Neanderthal World: 300,000–30,000 Years Ago

The Neanderthals inhabited northern Eurasia during the later phases of the Middle Pleistocene (300,000–130,000 years ago) and much of the succeeding Late Pleistocene (130,000–30,000 years ago). In contrast to their predecessors, they occupied the higher latitudes during a period when climates were generally cooler than today, although they endured at least one very warm episode at the beginning of the Late Pleistocene.

As in the case of the earlier Middle Pleistocene, the overall framework for climate change during Neanderthal times is provided by the stable-isotope record (see box 1 in chapter 3). In addition to cores from deep-sea deposits, records are also available from the Antarctic and Greenland ice sheets for the final

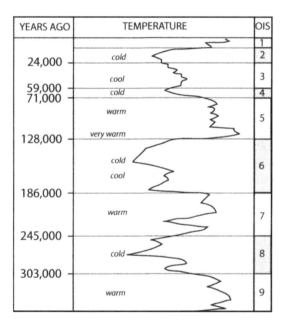

Figure B2. Temperature curve based on fluctuations in the ratio of oxygen isotopes from a deep-sea sediment core.

also retain some primitive traits and may be considered an early form of Neanderthal.[10] Skulls of comparable age from England (Swanscombe) and Germany *(Steinheim)* reveal a similar pattern.[11]

By 250,000–200,000 years ago, the full complement of Neanderthal traits was present. A skull from *Ehringsdorf* in Germany dated to roughly 230,000 years ago, and two skulls from *Biache* in France—dating to 190,000–160,000 phases of the Middle Pleistocene. The stable-isotope cores indicate that two major glacial intervals—oxygen-isotope stages 8 and 6 (OIS 8 & 6)—separated by a relatively cool interglacial (OIS 7), occurred between 300,000 and 130,000 years ago. In contrast to earlier periods, a number of terrestrial deposits can be correlated with these events, including long pollen cores, loess stratigraphy, and ancient cave sediments.[a]

The final glacial period of the Middle Pleistocene was followed by the Last Interglacial climatic optimum (OIS 5e) between 128,000 and 116,000 years ago. For at least a few millennia, mean annual temperatures climbed at least one or two degrees Centigrade above those of the present day, and much of Europe was covered by dense forest and marsh. Climates were relatively cool during the succeeding interval (OIS 5d–5a), but Neanderthal sites are especially common at this time. A typical setting may be found in the southwest region of the East European Plain. In the Dnestr Valley, where evidence of earlier settlement is rare, the Neanderthals occupied an open woodland dominated by pine. Temperatures were at least a few degrees cooler than those of today, and local mammals included woolly mammoth, horse, and steppe bison.[b]

Full glacial conditions prevailed during the Lower Pleniglacial (OIS 4), which is dated to between roughly 75,000 and 60,000 years ago. Milder but oscillating climates characterized the Middle Pleniglacial (see box 3 in chapter 5), and many sites are also dated to this final period of Neanderthal settlement.

---

a. N. J. Shackleton and N. D. Opdyke, "Oxygen Isotope and Paleomagnetic Stratigraphy of Equatorial Pacific Core V28-238: Temperatures and Ice Volumes on a 10^3 and 10^6 Year Scale," *Quaternary Research* 3 (1973): 39–55; G. Woillard, "Grande Pile Peat Bog: A Continuous Pollen Record for the Last 140,000 Years," *Quaternary Research* 9 (1978): 1–21; Henri Laville, Jean-Philippe Rigaud, and James Sackett, *Rock Shelters of the Perigord* (New York: Academic Press, 1980), pp. 144–215; J. Jouzel et al., "Extending the Vostok Ice-Core Record of Palaeoclimate to the Penultimate Glacial Period," *Nature* 364 (1993): 407–412.

b. Clive Gamble, *The Palaeolithic Settlement of Europe* (Cambridge: Cambridge University Press, 1986), pp. 160–176; John F. Hoffecker, *Desolate Landscapes: Ice-Age Settlement in Eastern Europe* (New Brunswick, NJ: Rutgers University Press, 2002), pp. 28–34.

---

years ago—exhibit virtually all of the features of the so-called classic Neanderthals of Western Europe. Most of the latter date to the last glacial cycle, which began 130,000 years ago.[12]

To the modern humans who eventually met up with them, the physical appearance of the Neanderthals must have seemed odd and perhaps grotesque. They possessed a long and low cranial vault with a low receding

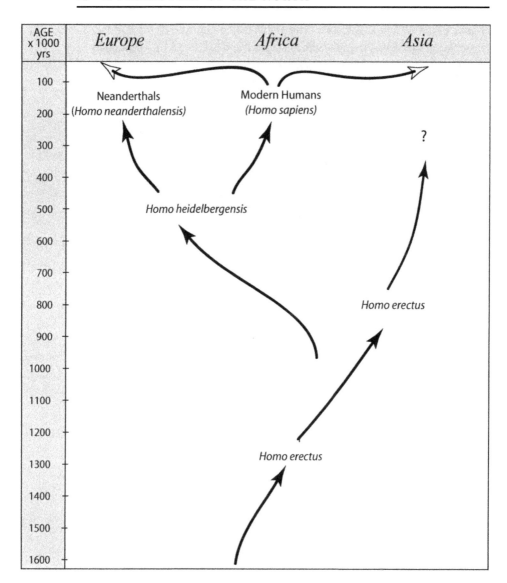

Figure 4.1. Evolution of the genus *Homo*, illustrating the relationship between Neanderthals and modern humans.

frontal bone and large brow ridge *(supraorbital torus)*. Their brain volume nevertheless was large—comparable to that of modern humans—averaging slightly more than 1,500 cc. The back of the cranium projected outward to form an *occipital bun*. The face projected forward with inflated cheeks and a large nasal cavity. The front teeth were exceptionally large relative to the cheek teeth, and the jaw lacked a chin.[13]

The Neanderthal skeleton below the neck was also quite distinctive. The ribs were large and weakly curved, creating a thick chest. The limb bones were robust with large areas for muscle and ligament attachments. The lower arm and upper leg bones were slightly bowed. The distal limb segments were short relative to the proximal segments. The tips of the fingers were large and rounded relative to those of modern humans.[14]

The emergence of the classic Neanderthals was accompanied by changes in technology. Broadly coincident with the skulls from Ehringsdorf and Biache are the earliest dated sites of the *Mousterian* industry (Middle Paleolithic). In contrast to the preceding Acheulean, hand axes and other large bifacial tools became rare or absent. And prepared-core techniques—allowing improved control over the size and shape of stone-tool blanks—became relatively common. For the first time, there is evidence for hafting flaked-stone pieces to handles and shafts, representing the oldest-known composite tools and weapons.[15]

Cold Weather People

The Neanderthals were a northern form of human in the same way that the arctic hare is a northern form of the jackrabbit. They evolved the most extreme anatomical adaptations to cold climates ever found among hominids, and have been characterized as "hyperpolar." Both the size and shape of their bodies acted to minimize heat loss and the danger of cold injury. They were stocky and barrel-chested with large heads and short limbs. And like living peoples of the circumpolar zone, they almost certainly evolved some physiological mechanisms for warming the extremities and exposed surface areas.[16]

It is possible to overstate the extent to which Neanderthal anatomy reflects cold adaptation. The large size of their heads has been mentioned as a possible example, but—as in the case of *Homo heidelbergensis*—an increase in brain size took place in both Africa and Europe at this time and again may have had little to do with climate.[17] As already noted, many Neanderthal traits seem to have been the result of isolation and random genetic drift. Other features, such as the large front teeth, may reflect an evolutionary response to aspects of their environment other than low temperature.

Nevertheless, many features of Neanderthal anatomy—especially postcranial anatomy—are almost certainly part of an adaptation to cold. The pattern was first noted by the American anthropologist Carleton Coon in a 1962 book entitled *The Origin of Races*. Coon compared many Neanderthal traits with the anatomy of living peoples at high latitudes. He observed that the size and shape of their bodies would have minimized the loss of heat

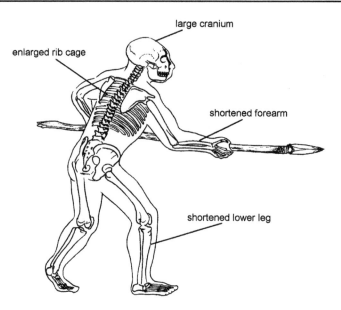

Figure 4.2. The Neanderthal skeleton, illustrating features interpreted as adaptation to cold climates.

and provided protection against cold injury. But by the time his book was published in late 1962, Coon had become embroiled in a bitter controversy concerning his views on race. As a result, his ideas about cold-climate adaptation among the Neanderthals were largely ignored or forgotten for many years.[18] Renewed study of the subject since 1980 has shown that most of his observations were valid.

Neanderthal limbs were short relative to trunk length. Their lower limb bones (radius and tibia) were especially short in comparison with the upper limb bones (humerus and femur). Among living human populations, the ratios of lower- and upper-limb segments correlate strongly with temperature and latitude. Significantly, the Neanderthals exhibit even lower ratios than those of the modern Inuit. Coon stressed the size and thickness of the foot bones as another means of conserving body heat and protecting against frostbite.[19]

The large overall body mass of the muscular Neanderthals further reduced the ratio of exposed surface area to volume. The thickness of the chest is particularly striking, and is reflected in the shape of the ribs and the length of the collar bone (clavicle).[20] A thick coat of body hair would have enhanced heat conservation, but it is currently impossible to confirm its presence.

Coon speculated that the Neanderthals also had evolved physiological means for warming their extremities and exposed surface areas. He drew attention specifically to the large size of the *infraorbital foramina*—openings on each side of the maxilla below the eye socket that allow blood flow to the cheeks—and noted a parallel with the Greenland Inuit. As among the latter and other modern Arctic peoples, elevated blood flow to extremities probably helped the Neanderthals avoid cold injury.[21]

More controversial was Coon's suggestion that the large size of the nasal cavity was another cold-climate adaptation. Citing medical research on the vulnerability of the brain to very cold air, Coon proposed that the voluminous Neanderthal nose had functioned like a radiator to warm inhaled air. The "radiator nose theory" was critiqued by some prominent anthropologists, but may retain some validity. The nose might also have functioned to expel warm air and reduce excess body heat generated by intense activity.[22]

The unusual chewing complex of the Neanderthals has provoked comment as well. In addition to the forward position of the jaws and the large size of the front teeth, the latter betray extreme wear in all but the youngest individuals. Furthermore, the wear includes microscopic traces of nonfood items. The pattern suggests that the Neanderthals were using their teeth to grip—and perhaps to process—hide and other soft materials. Here again there is a parallel with the Inuit, whose teeth often exhibit similar wear from such use.[23]

The powerful muscles of the Neanderthals, which are indicated by the deep muscle attachments on their bones, are widely thought to reflect a stressful life and a greater reliance on brawn rather than on technology. Further evidence of physical power and stress may be seen in the morphology of the shoulder blade (scapula), thick-walled limb bones, and other features.[24] Though not cold-climate adaptations in themselves, Neanderthal strength and endurance were probably an indirect measure of their response to the glacial environments of Europe—especially in light of their limited technological achievements.

Neanderthal Geography in Space and Time

The geographic distribution of their sites provides one of the most important clues to Neanderthal tolerance for cold environments. Both the presence and the absence of sites in specific areas and time periods tell us much about the Neanderthals' ability to cope with low temperatures and other features of such environments. Overall, it seems that they were able to inhabit places where mean winter temperatures fell significantly below the freezing point—probably as low as 5° F (−15° C) and possibly −5° F (−20° C). They were unable to occupy the extreme cold regions of Northeast Asia where winter temperatures often average below these values today.

The occupation of Western Europe was continuous throughout the warmest and coldest fluctuations of the last glacial cycle. Neanderthal sites are especially numerous in southwest France and northern Spain (Franco-Cantabrian area), where conditions during glacial phases were ameliorated by the oceanic effect. Long sequences of occupation levels have been excavated at French caves and rock shelters like *Pech de l'Azé* and *Combe Grenal*. Even during the coldest intervals, mean January temperature probably fell no more than a few degrees below the freezing point. Trees were always present, although at times they were chiefly confined to scattered pine. During these intervals, reindeer dominated among large mammals in the region.[25]

In fact, Neanderthal settlement in Western Europe may have been more constrained during the warmest period of the last few hundred thousand years. The latter took place between 128,000 and 116,000 years ago and is usually referred to as the *Last Interglacial* climatic optimum. Mean annual temperatures were a couple of degrees higher than today, and hippopotamus *(Hippopotamus amphibius)* was living as far north as England. Evidence for human occupation in the Franco-Cantabrian area during the climatic optimum is very limited (though at some sites, occupation debris may have been washed away by an episode of intense erosion that followed this period).[26] The Neanderthals apparently were not partial to the humid forests and swamps of Western Europe at this time.

By contrast, occupation during the coldest phase, which dates to roughly 75,000–60,000 years ago—often termed the *Lower Pleniglacial*—is well documented at Pech de l'Azé, Combe Grenal, and other caves in southwest France (occupations seem to be much less common in northern France). Among the stone artifacts and mammal bones lie frost-shattered rock fragments from the cave walls in sediments that are rich in unweathered carbonates but poor in clays. These testify to a period of peak cold and aridity in the region.[27]

The extraction and analysis of fossil DNA from a specimen in the northern Caucasus Mountains in 2000 provided a remarkable opportunity to look at genetic variability *within* the Neanderthals. Comparison with the fossil DNA recovered earlier from the Neander Valley specimen allowed an estimate of the time of divergence between the West and East European populations of between roughly 350,000 and 150,000 years ago.[28] With a midpoint of 250,000 years, this estimate may be broadly accurate, although most of the early Mousterian sites and Neanderthal fossils in Central and Eastern Europe seem to date to the Last Interglacial (about 125,000 years ago).[29]

Perhaps the exceptionally warm conditions of the Last Interglacial were a trigger for expansion eastward into the cooler and drier parts of Europe.

Figure 4.3. Map of Neanderthal sites in Europe and Asia.

Nevertheless—even under interglacial climates warmer than today—estimated mean January temperatures in the areas of the East European Plain occupied at this time are 26° F (−3° C) or lower.[30] Both the morphology and the diet of the Neanderthals were probably prerequisites to occupation of this region.

In the more continental setting of the East European Plain, there may actually be a reversal of the pattern found in the Franco-Cantabrian region. Last Interglacial occupations seem to be more common than those of the cooler periods that followed. The site of *Khotylevo* on the Desna River (roughly 200 miles or 350 km southwest of Moscow) may be typical. This site actually comprises a series of artifact concentrations along a half-mile stretch of the ancient floodplain. An open woodland environment existed—warmer than today—inhabited by mammoth, red deer, and bison.[31]

Neanderthal settlement history in Eastern Europe varies dramatically through time. Much of the plain was probably abandoned during the extreme cold of the Lower Pleniglacial. At that time, mean January temperatures would have fallen to −15° F (−26° C) or lower, which seems to have exceeded the cold tolerance of the Neanderthals. Some areas may have been reoccupied during the milder interstadial period that began after 60,000 years ago. Along the southernmost margin of Eastern Europe—where conditions were closer to those of Western Europe—occupation seems to have been continuous.[32]

During the Last Interglacial climatic optimum, the Neanderthals moved even further eastward into southernmost Siberia. The small cave of *Dvuglazka,* situated near the town of Abakan on a tributary of the Yenisei River, contains evidence for Neanderthal occupation at this time. Although Dvuglazka lies at roughly the same latitude as Khotylevo, local climates are much harsher and the current January mean is 3° F (−17° C). The lack of cold-loving mammals in the layers apparently occupied by Neanderthals suggests that conditions were milder during the climatic optimum. No Neanderthal skeletal remains were recovered from the cave, but typical Mousterian artifacts were found along with the mammal bones.[33]

Equally impressive are a group of caves and open-air sites located southwest of Dvuglazka in the Altai region at latitude 51° North. Conditions are somewhat milder here, but occupation seems to have continued after the Last Interglacial into much cooler periods—possibly including the Lower Pleniglacial. Isolated skeletal remains were found in two of the caves and are tentatively identified as Neanderthal. In many occupation layers, the mammal bones include arctic taxa such as reindeer and polar fox.[34] Mean January temperatures may have fallen as low as −5° F (−20° C), and the Altai sites appear to represent the maximum cold tolerance of the Neanderthals.

One of the most intriguing aspects of Neanderthal geography is their southward expansion into Central Asia and the Near East. This seems to have taken place during the Lower Pleniglacial, when some areas of northern Eurasia occupied during the Last Interglacial were apparently abandoned. At that time—and possibly during other intervals as well—Neanderthals moved as far south as latitude 33° North in the Levant. Their sites are also found in northern Syria *(Dederiyeh Cave)* and Iraq *(Shanidar Cave)* and further north and east in southern Uzbekistan *(Teshik Tash)*.[35] Significantly, the anatomy of the southern Neanderthals reveals a less extreme adaptation to cold climates—lower-limb proportions are increased while the thickness of the chest is reduced.[36]

New Technology

When Carleton Coon described the cold adaptations of the Neanderthals in 1962, he framed his observations with a provocative hypothesis. Surveying the archaeological record and noting the lack of evidence for any technological innovation, Coon suggested that the Neanderthals' adaptation was based primarily on anatomy and diet.[37] This became a recurrent theme in Neanderthal studies and one that implied a profound contrast with modern humans, who depended heavily on technology to survive in the same environmental settings.

In recent years, it has become apparent that the emergence and expansion of the Neanderthals was accompanied by some novel technological developments. The most important of these was the design and use of composite tools and weapons. By attaching stone blades and points to handles and shafts, the Neanderthals increased their power and efficiency. There may also have been new developments in the uses of wood and hide, but these are more difficult to specify at present. What role did the new technology play in Neanderthal adaptation and how critical was it to the occupation of cold habitat?

Archaeologists have traditionally focused their attention on the stone tool technology, emphasizing the disappearance of hand axes and rise of prepared-core techniques. The latter spread across the same regions where the hand ax industries had existed—west of the Movius Line—and both may be linked to the use of composite tools. First of all, hand axes were held and used in the hand, and the invention of composite tools must have reduced the need for them. Second, composite tools would have demanded increased control over the size and shape of stone blanks in order to accommodate wooden handles and shafts. This may have been the primary catalyst for the widespread adoption of prepared-core techniques.[38]

Although prepared-core (or *Levallois*) techniques appeared in late Acheulean sites roughly half a million years ago, they did not become common until the Mousterian. The techniques are relatively complicated and require a series of sequential steps as the stone is shaped and trimmed for the removal of a blank of predetermined size and shape. Among the Levallois variations is one that allows production of a sharp point—suitable for use as a spear tip. For reasons that remain obscure, the Neanderthals did not always choose to produce blanks with prepared cores. Many of their sites contain non-Levallois cores.[39]

Despite their complexity, prepared-core techniques do not seem to have called for any conceptual abilities beyond those already required for making hand axes.[40] Composite tools, in contrast, may reflect a significant advance in cognitive skill. Not only does the structure of the finished tool bear little resemblance to the raw materials from which it is fashioned, but at least three different elements must be brought together to create the form. These include the wooden shaft or handle, a stone blade, and an adhesive.[41]

No composite tools or weapons have been found in Neanderthal sites, but their existence is well documented. Both microscopic wear patterns and traces of adhesive (for example, bitumen) on stone artifacts indicate that they were secured in wooden hafts. Although the percentage of such artifacts is low, archaeologists suspect—on the basis of experimental studies—that both forms of evidence are rarely preserved and that the actual numbers of composite implements are higher. They apparently included woodworking tools—scrapers hafted to the sides and ends of handles—as well as stone-tipped spears.[42]

Hand axes and other large bifaces are rare in Neanderthal sites, but small bifaces are not uncommon in some areas. In general, these tools are better known from Central and Eastern Europe. The Neanderthals also produced a variety of tools from stone flakes that were held and used by hand. The backs of these tools were sometimes blunted for easier handling. Although archaeologists have defined more than sixty types of Mousterian flake tools, some types probably represent different stages of resharpening the same basic tool form. Overall the flake tools of the Neanderthals do not seem to have differed much from those of their predecessors.[43]

More interesting are the wooden tools and weapons. As in the case of the earlier European sites (see chapter 3), isolated examples of wooden artifacts have been found. A spear is known from the site of *Lehringen* (Germany), and wooden objects from *Abric Romani* (Spain) include a scooplike implement and possible clubs.[44] These discoveries of wooden artifacts—

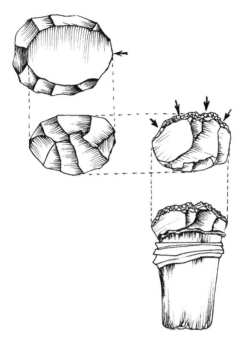

Figure 4.4. Prepared (or Levallois) core and flake and reconstruction of a composite flake tool.

so rarely preserved—combined with abundant evidence of woodworking from microwear studies of stone tools,[45] suggest that wood technology might have been relatively diverse and complex. But the extent to which the Neanderthals had improved and expanded such technology over *Homo heidelbergensis* remains unclear.

Among recent peoples of the Arctic, clothing and shelter seem to be the most complex aspect of technology.[46] They have always been a vital component of modern human adaptation to high latitudes. Here may lie the most significant contrast between modern humans and Neanderthals—at least with respect to cold adaptations—and the critical constraint on the ability of the latter to cope with extreme cold. Although often difficult to evaluate, the evidence for Neanderthal clothing and shelter technology generally indicates a low level of complexity and effectiveness.

Much of it is negative evidence. Some stone tools exhibit microwear polish from hide-working, but the polish reflects only the initial stage of hide preparation. The complete absence of sewing needles is especially important. Among preindustrial modern humans, these implements are invariably fashioned out of bone or ivory and should be preserved in sites where animal remains have not been heavily weathered (and where archaeologists sieve excavated sediment for recovery of tiny objects). But stone and bone awls—though rare—are reported from some sites, and microwear polish on these reveals hide-working.[47]

Pieced together, these bits of information suggest that the Neanderthals may have been using hides for protection against low temperatures, but with only minimal modification. Perhaps they were employed primarily as blankets. The awls indicate that sometimes hides may have been perforated and bound with line. The lack of needles, however, shows that they were not cut and sewn into tailored clothing—providing effective insulation in very cold weather, but without restricting the ability to move and perform activities.[48]

Evidence for the construction of artificial shelters is almost unknown. In this case, the most significant information is to be found on the East European Plain, where caves and rock shelters are rare. Most sites in this vast region are open-air locations on river terraces. When modern humans later occupied Eastern Europe, they left numerous traces of huts with interior hearths. But the Neanderthal sites on the plain lack any convincing remains of former structures, although several sites contain patterns of mammoth bones that might represent a simple windbreak.[49]

The 1982 French film *Quest for Fire* entertained audiences with the drama of a Neanderthal group attempting and eventually failing to preserve a flame. It was based on the intriguing idea that the Neanderthals never developed the technology for producing fire and relied on finding and maintaining a natural flame. Although the idea cannot be confirmed, it should be noted that several recent hunter-gatherer peoples reportedly lacked the ability to make fire—a fact that underscores both the complexity of this technology and the possibility of living without it.[50] In any case, both Neanderthal caves and open-air sites contain abundant evidence for the use of controlled fire in the form of former hearths.[51]

A Society of Hunters

The diet of the Neanderthals was probably as least as critical an adaptation to cold environments as their anatomy. Without a diet high in protein and fat, it is difficult to imagine how they could have sustained themselves across most of their range. This is especially true in the coldest and driest regions—Eastern Europe and southern Siberia—but may have applied to the milder parts of that range as well. The Neanderthals were probably the first hominids to consume a diet composed primarily of meat.

The most important information on Neanderthal diet is derived from the chemical analysis of their bones. Stable-isotope values for specimens from *Scladina Cave* (Belgium) and *Vindija Cave* (Croatia) are comparable to those of various northern carnivores and suggest a predominantly meat diet. Especially striking are the results from Scladina Cave—where the deposits date to the Last Interglacial climatic optimum, and climates were warmer than today and plant foods abundant.[52]

The Neanderthals probably had higher meat requirements than modern humans in comparable environmental settings. Like living peoples in northern latitudes, they must have had a high basal metabolic rate and greater caloric needs. And because the Neanderthals had evolved a large body mass and exceptionally powerful muscles, their energy needs must have been especially high. Their lack of well-insulated clothing would have further increased the need for high caloric intake in low temperatures.[53]

Most of the meat consumed by the Neanderthals seems to have been obtained from large mammals. The bones and teeth of a variety of species are found—often in large quantities—in cave and open-air sites throughout the Neanderthal range. In Western Europe, woodland forms such as red deer are common, although reindeer becomes numerous during the colder periods. In Eastern Europe, where open steppe habitat was more widespread, species like bison and saiga are more typical. In upland areas such as the Northern Caucasus, the local Neanderthals sometimes ate sheep and goat.[54]

Most of the large-mammal meat consumed seems to have been obtained by hunting. Animal bones recovered from Neanderthal sites often exhibit traces of hammer blows and incisions from stone tools, indicating intensive butchering of carcasses. Among the adult mammal remains at these sites, prime-age individuals frequently dominate—a pattern not found in remains accumulated by scavenging. In the case of the latter, old adults normally dominate the assemblage, reflecting natural deaths among the animal population. In fact, it is more difficult to find evidence of scavenging than hunting in Neanderthal sites.[55]

Although effective hunters not only of reindeer and goat but also of much larger prey such as red deer, horse, and bison, the Neanderthals may have found it difficult to kill mammoths. Mammoth bones and tusks often turn up in their sites, but usually in modest numbers. The bones sometimes exhibit a more weathered appearance than the other remains, along with traces of gnawing by large carnivores. At least some of these bones and tusks seem to have been gathered from natural occurrences and brought to the site, perhaps—as already noted—for use as windbreaks. A rare example of mammoth hunting may be at *La Cotte de St. Brelade* on Jersey (Channel Islands off the coast of France).[56]

At the other end of the size spectrum, small-mammal and bird remains also show up in many Neanderthal sites, but—for the most part—these species do not appear to have been food debris. The bones seem to have found their way into the cave or open-air location by other means. Most of them represent either small animals that nested in the same caves or open-

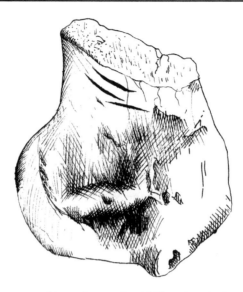

Figure 4.5. Large mammal bone fragment exhibiting stone-tool cut marks recovered from a Neanderthal site.

air locations occupied by Neanderthals or bones collected by other animals. Although there is some evidence for shellfish gathering along the Mediterranean coast (for example, at *Grotta dei Moscerini* in Italy), fish remains are almost entirely absent and stable-isotope analyses of Neanderthal bones indicates that fish were not an important part of the diet.[57]

The apparent absence of small game in the diet reveals a significant contrast with modern humans. Both the people who succeeded the Neanderthals in northern Eurasia and more recent inhabitants of higher latitudes typically supplemented their large-mammal diet with a variety of small mammals and birds, while fish is an important dietary component among virtually all northern foraging peoples. The contrast would seem to reflect the inability of the Neanderthals to design the more intricate forms of technology—traps, snares, weirs, nets, throwing darts, and so forth—that hominids require to catch these less accessible and more elusive prey.[58]

All of this emphasis on the hunting of large mammals and limited reliance on scavenging, plant gathering, and small-game exploitation has profound implications for the way in which the Neanderthals distributed themselves across the landscape. Large mammals are highly mobile and in order to ensure a steady food supply, their movements must be followed and/or anticipated. Because it is difficult for humans to keep up with a herd, interception is the more practical and common strategy.

One of the best areas to view the distribution of Neanderthal sites on a regional scale is the western end of the Northern Caucasus. In this area, many years of archaeological survey and excavation have yielded a mass of information on cave and open-air sites located at varying elevations on the mountain slope. Sites range from those in the lower foothills at only about 300 ft (100 ms) above modern sea level to others at elevations of 2,300 ft (720 ms) and *Mezmaiskaya Cave* at 4,200 ft (1,300 ms). Although data on the seasons during which each of these sites was occupied remain sketchy, it appears that the Neanderthals were using different sites at specific times of the year to hunt seasonally available mammals. Bison were probably hunted in the foothills in the late fall and winter; goat and sheep were killed during the summer months at higher elevations.[59]

At most of these sites, Neanderthals were retrieving the carcasses or parts of carcasses from other (presumably nearby) locations. This is true at both the caves and open-air sites in the Northern Caucasus and elsewhere—it is otherwise impossible to account for the massive accumulations of large mammal bone in these sites. Although there may be uncertainties and ambiguities about the practices of their European predecessors *(Homo heidelbergensis)*, there can be no doubt that the Neanderthals were central-place foragers (and this is fully consistent with the predictions of foraging theory among ecologists).[60]

The pattern observed in the Northern Caucasus indicates that Neanderthal groups were scheduling their movements to take advantage of seasonal concentrations of large-mammal prey. Along with their meat diet, this

Figure 4.6. Mezmaiskaya Cave on the northern slope of the Caucasus Mountains.

may have been a prerequisite for survival in the colder parts of their range. In any case, the pattern suggests at least some degree of coordination and planning, and raises a fundamental and contentious issue in Neanderthal studies. Modern humans plan and coordinate their activities with the use of spoken language, which allows them to create and communicate abstract models of the world.

Did the Neanderthals also possess language and, if not, how did they plan and coordinate their foraging movements? One of their minor—but potentially important—anatomical differences with modern humans was the position of their vocal tract. The shape of the *basicranium* (or base of the skull) indicates that the larynx was located higher in the neck than it is among ourselves. This would have restricted the Neanderthals to a smaller range of vocal sounds and denied them the power of rapid speech.[61] Slower and more limited speech, however, by no means precludes some form of language.[62]

More significant perhaps is the scarcity of evidence for symbols in the Neanderthal archaeological record. Language is only one of the many ways by which modern humans structure and model the world with symbols. As described in the next chapter, material traces of other expressions of symbolism in the form of art objects, musical instruments, ornaments, and so forth are common in modern human sites after 50,000 years ago. These artifacts—along with evidence for a shift in the position of the vocal tract—help confirm the presence of symbolic language among modern humans at that time.[63]

But similar material expressions of symbolism are virtually unknown in the sites of the Neanderthals, and this absence helps reinforce the suspicion that they were without language as we know it. Simple ornaments and art objects have been reported from several sites, but they seem to be extremely rare at best. In some cases, the association with Neanderthal remains is uncertain.[64] A punctured bear bone recovered in 1997 from a cave in Slovenia and interpreted as a flute now appears to be a fragment chewed by carnivores.[65]

Burial of the dead may fall in a different category. Intentional burials have been reported from Neanderthal caves and rock shelters since 1908 and have always been somewhat controversial. Sometimes "grave goods" in the form of animal remains or stone artifacts have been reported from these burials.[66] After decades of often rancorous debate, most archaeologists accept the burials as genuine, although many question the interpretation of the grave goods.[67] Debate continues, however, over the motive for burial. Some believe that the Neanderthals were merely disposing of corpses as undesirable waste, but many find this explanation unconvincing.[68]

The graves are a mystery and the most important clue that the Neanderthals probably were not as simple as they appear. Intentional burial would seem to reflect some concept or set of ideas regarding death and possibly afterlife. Such behavior is unknown among the living apes. The lack of supporting evidence for symbolism in the form of art, music, and ornamentation suggests that these concepts might have been structured and expressed in a manner very different from those of modern humans.

The apparent absence of language and symbolism also carries implications for the organization of Neanderthal society. Among modern humans, organization outside the immediate family is based primarily on shared symbols. It is difficult to imagine social life among any living peoples without a colorful array of clothing styles, table manners, religious rituals, jokes, and so forth. The seeming absence of such things among the Neanderthals raises questions regarding how they held their societies together.[69]

Perhaps Neanderthal social life was based on their communication system, combined with simple customs such as burial of the dead. Although limited in comparison with modern humans, their speech—accompanied by a wide range of gestures—and various customs (most of which may not have preserved well in the archaeological record) may have provided a basis for establishing relationships among groups beyond the immediate family. Alternatively, their society might have been organized along radically different lines than that of modern humans, and there has been much speculation—some of it rather bizarre—about the nature of Neanderthal organization. One archaeologist suggested that adult males and females lived in separate groups and met only periodically to mate.[70]

The heavy emphasis on the hunting of large mammals introduces some likely constraints to social organization. Among recent foraging peoples, the sexual division of labor increases significantly as the percentage of gathered food declines and males assume a proportionally greater role in procuring meals. As in modern humans, it seems unlikely that pregnant or nursing females played a major role in hunting large mammals, and male-female pair bonds were probably a cornerstone of Neanderthal society.[71]

However, the number of male-female pairs and their offspring in a typical Neanderthal group may have been smaller than that found among recent foraging peoples. Estimating the size of a group of people who occupied a given archaeological site is difficult, because it is rarely possible to be certain that the entire area of the site was used at one time. Most sites probably were used repeatedly by one or more groups and are composed of a mixture of debris that is impossible to sort out in terms of specific groups. Nevertheless, in a few cases it is possible to at least define the maximum limits of the habitation area. At sites like *Grotte du Lazaret* (southern France) and *Barakaevskaya Cave* (Northern Caucasus), the area used appears quite small—less than 360 ft^2 (or about 40 m^2)—suggesting a group of no more than a dozen people.[72] By comparison, residential groups among recent foraging peoples typically number about twenty-five individuals.[73]

Of course, the examples mentioned above might be atypical, and larger groups could have occupied many sites. But the evidence for long-range movements of Neanderthal groups is entirely consistent with the notion of smaller societies. If the distance traveled by raw materials from their sources—such as flint used for tools—accurately reflects the movement of the people who carried them, the Neanderthals were moving around in smaller ranges than those of recent foragers in similar environments. Although some isolated exceptions are known, Neanderthal sites rarely contain raw materials carried more than 60 miles (100 km) from their source area. Smaller ranges and territories would have supported smaller groups.[74]

What Went Wrong? The Fate of the Neanderthals

The Neanderthals disappeared from the fossil record approximately 30,000 years ago. Their disappearance—and the more or less simultaneous appearance of modern humans—is one of the more contentious issues in paleoanthropology.[75] There is no reason to believe that the Neanderthals became extinct because they could not adapt to changing environments at this time. Although there is evidence for abandonment of the coldest parts of their range during the Lower Pleniglacial, by 50,000 years ago they were on the rebound—possibly reoccupying previously vacated areas.

Carleton Coon, who provided so many insights into the Neanderthals, was one of a number of anthropologists who believed that they evolved directly into modern humans.[76] This thesis now seems more than improbable, given the striking anatomical differences between the two forms (supported by the genetic distance indicated by fossil DNA studies). Moreover, the modern humans who succeeded the European Neanderthals exhibit anatomical features associated with warm climates, and it is difficult to explain why they would have evolved such features in the midst of the last glacial period.[77]

Instead, it seems all but certain that the disappearance of the Neanderthals was closely tied to the arrival of modern humans from southern latitudes. Debate rages, however, over the process by which the transition took place. Many argue that it was essentially one of gene flow from Africa, and that the modern humans of northern Eurasia ended up with a mixed inheritance.[78] A modern human skeleton excavated in 1998 from *Lagar Velho* in Portugal exhibits some Neanderthal features (for example, short distal limb segments) and is offered as an example of interbreeding between the two populations.[79]

Others are highly skeptical of this scenario and argue that outright replacement of Neanderthals by modern humans—with little or no genic exchange—is more likely.[80] The challenge to adherents of this view is explaining why the Neanderthals, who had adapted with such obvious success to the glacial environments of northwestern Eurasia, were overwhelmed by people who were so ill suited to life in these environments.

CHAPTER 5

Modern Humans in the North

The Revolution

Modern humans evolved gradually in Africa and later dispersed—rather suddenly—into Eurasia and Australia. Between roughly 60,000 and 40,000 years ago, they expanded the human geographic range into new habitats and regions in both the southern and the northern latitudes. This dramatic expansion and rapid adaptation to a variety of settings was effected by a behavioral revolution related to spoken language and other forms of symbolism. Humans acquired an unprecedented ability to manipulate the surrounding world and used it to create a place for themselves almost everywhere.

Although the habitats occupied in the southern latitudes varied from rain forest to desert margin, without a doubt the most impressive aspect of the expansion out of Africa was the invasion of northern Eurasia. Despite their recent tropical origins, modern humans spread quickly across the cold steppes and forests of Europe and Siberia. Between 45,000 and 24,000 years ago, they settled areas up to latitude 60° North and possibly further at times. During this period, climates in northern Eurasia were colder than at present, and winter temperatures in some of these areas probably fell below anything endured by earlier humans.[1]

Modern humans achieved their expansion into new cold environments almost entirely through novel behavior. As noted briefly at the end of the preceding chapter, they arrived in northern Eurasia with anatomical features better suited to the equatorial zone. Yet they managed to occupy colder places than those long inhabited by the cold-adapted Neanderthals. The archaeological record reveals that this was accomplished largely through technological innovation.[2]

Much of the novel technology documented or inferred from their sites at this time may be found among ethnographic accounts of recent foraging peoples of the Subarctic and Arctic. It included tailored fur clothing, portable lamps, and heated shelters for protection from extreme low temperatures. It also included implements for snaring and trapping small game and for catching birds and fish. Modern humans in northern Eurasia consumed a high protein-fat diet like the Neanderthals, but they designed new instruments to expand the range of their food sources. The novel technology included fired ceramics, woven textiles, and possibly the first mechanical devices.[3]

The creative abilities that underlay all this innovation and complexity in technology were linked to spoken language and other forms of symbolism that became so evident in the archaeological record after 100,000 years ago. By all indications, modern humans had begun at this time to classify and structure everything around them in words, visual art, musical sounds, and other symbolic media. As they arranged and recombined symbols to create new sentences, songs, sculptures, and other forms, they created equally novel implements and devices.[4] In fact, spoken language was in itself a powerful device for invading new habitats—providing a ready means for classification of strange plants and animals.

After 28,000 years ago, glaciers in the northern hemisphere began to expand and climates became increasingly cold and arid. This period is known as the *Upper Pleniglacial*, and it effectively brought arctic conditions to the middle latitudes for several thousand years. During the Upper Pleniglacial, modern humans living in northern Eurasia developed many of the same adaptations later observed among the Inuit and other foraging peoples of the Arctic. However, as cold climates approached their peak intensity roughly 24,000 years ago, people abruptly abandoned large portions of the East European and Siberian lowlands.[5]

Retreat from the coldest parts of their range brought modern humans back to roughly the same geographic distribution enjoyed by the Neanderthals during the preceding glacial interval (the Lower Pleniglacial). The reason why modern humans were unable to inhabit these parts during the cold peak is by no means obvious. Equipped with similar technology, recent foraging peoples like the *Yukaghir* in Northeast Asia coped successfully with equally severe climates. The answer seems to lie in their tropical anatomy and high susceptibility to cold injury under these conditions—regardless of clothing and shelter.[6]

Modern Human Origins

In many ways the evolutionary origins of modern humans parallel those of the Neanderthals. Modern humans also evolved gradually from *Homo heidelbergensis* during the same broad interval of time (roughly 600,000 to 200,000 years ago). However, the specialized features that developed among the Neanderthals—many of them in response to cold climates in Europe—are absent in the African fossil record.

Characteristic *Homo sapiens* traits may be seen in African fossils between 300,000 and 150,000 years ago. As the overall robusticity of the skeleton decreased, the size of the brain became larger. A partial skull from *Florisbad* (South Africa), dated to approximately 250,000 years ago, exhibits the trend toward a more vertical forehead (frontal bone) and less projecting face. Younger remains from Morocco *(Jebel Irhoud)*, Sudan *(Singa)*, and Ethiopia *(Omo)*, while retaining a few archaic features (for example, a thick brow ridge), are essentially modern in appearance.[7]

The altered shape and increased size of the brain may reflect important changes in its function—thought by some to represent the emerging "executive brain." The executive functions reside in the frontal lobes of the brain. They include decision-making abilities, organization and planning, sequential memory, and the ability to think about the past and the future.[8]

Along with the changes in brain anatomy, the most significant evolutionary development in Africa seems to have been the position of the vocal tract. Analyses of the shape of the base of the skull (basicranium) suggest that the larynx had shifted downward into the modern human position by 100,000 years ago, if not earlier. This suggests that early *Homo sapiens* had developed the anatomical capabilities for fully modern speech. The lowered position of the larynx increases the danger of choking on food—a problem that remains unsolved among modern humans—underscoring the importance of this change and the strong selection pressures that lay behind it.[9]

If much of Neanderthal evolution may be explained as a response to cold climates, it is more difficult to explain why African *Homo heidelbergensis* evolved into modern humans. Climate would seem to have played little or no role in *Homo sapiens* origins, and the African fossils—as might be expected—continue to exhibit the same warm-climate adaptations first noted in *Homo erectus* (see chapter 2).[10] The increase in brain size and other changes in cranial anatomy are probably tied in some way to the evolution of the vocal tract and the behavioral changes associated with language.

Why is human language so important and unique? Derek Bickerton—one of the few professional linguists to study the origins of language—argues that only humans have incorporated a complex "representational" system into their means of communication. Virtually all animals have evolved representational systems in their brains for mapping or modeling the environment. For example, domestic dogs and cats generate mental maps of the houses that they inhabit and do not have to relearn the locations of and access routes to specific objects (such as their food bowl) each time they want or need them.[11]

Most animals—especially those living in social groups—also have evolved means of communication. Animal communication systems, however, carry very simple bits of information (for example, an alarm call or a threat display). According to Bickerton, language appeared when humans evolved the neural mechanisms and vocal apparatus to communicate mental mod-

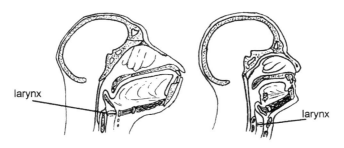

Figure 5.1. Evolution of the vocal tract from apes (left) to modern humans (right).

els of the environment by generating coded sequences of sound. By combining and recombining sounds into words and words into sentences, humans could create an almost infinite variety of meanings.[12] Modern humans also developed the ability to transmit mental models through nonverbal means such as visual art and music. Moreover, by communicating their mental models to each other, humans made them common property among social groups.

The timing of the behavioral changes that evolved with modern humans is a matter of debate. In the fall of 2000 Alison Brooks and Sally McBrearty published a lengthy paper in the *Journal of Human Evolution,* arguing that the use of symbols and new technologies developed slowly between 300,000 and 50,000 years ago. Many of the archaeological clues associated with language and modern human behavior appear in Africa during this interval. They include artifacts made on stone blades, use of pigment, long-distance movement of raw materials, bone tools and weapons, and possibly some ornament and art.[13] This view fits nicely with the anatomical evidence for an early shift in the position of the vocal tract.

A radical alternative has been put forward by Richard Klein, who believes that language and modern human behavior evolved suddenly about 60,000–50,000 years ago. According to the "neural hypothesis," fully modern speech was the result of a random genetic mutation that spread quickly among African *Homo sapiens* at that time.[14] Bickerton suggests that the crucial mutation was related to the appearance of syntax—the complex rules that govern sentence formation in all languages.[15] Biomolecular evidence for such a mutation has emerged recently with the discovery that a gene linked to speech function (designated "FOXP2") evolved into its current form less than 200,000 years ago.[16]

The dating of much of the African evidence for the use of symbols and new technologies prior to 60,000–50,000 years ago is admittedly problematic. For example, reported bone harpoons from *Katanda* (Zaire) are dated to 150,000–90,000 years, but may actually be younger than 25,000 years.[17]

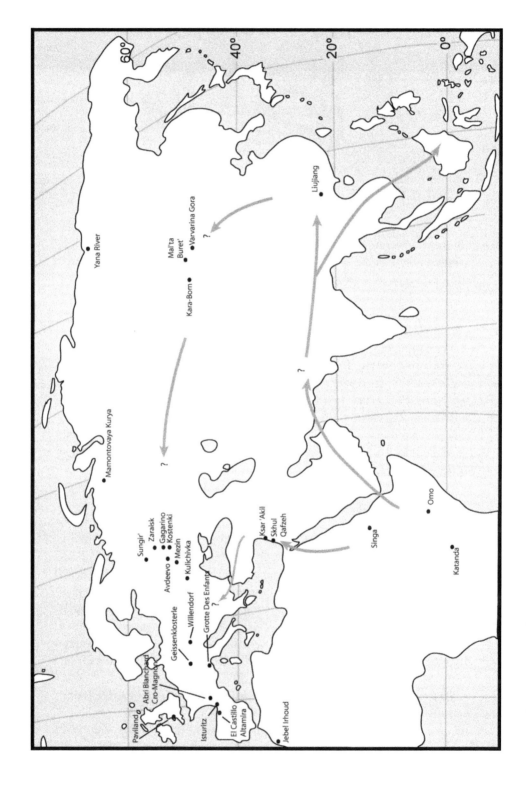

Recently, both engraved fragments of ochre and bone tools were found at *Blombos Cave* (South Africa) in deposits that date to approximately 70,000 years.[18] Although the dates on the bone tools and art from Blombos Cave are widely accepted, archaeologists continue to debate this issue.

In any case, it appears that language and the use of symbols had reached a level comparable to that of recent and living peoples by the time that *Homo sapiens* began to disperse out of Africa roughly 60,000–50,000 years ago. From that time onward, the archaeological record yields increasing evidence of complex technology unknown among earlier humans. Whether fully modern human behavior evolved gradually over a period of a quarter of a million years or immediately prior to the dispersal, both the use of symbols and the capacity for innovative technology seem to have a reached a threshold that may have been critical to the rapid expansion into Eurasia and Australia.

Out of Africa Again

The dispersal of modern humans out of Africa was an event remarkable not only in hominid evolution but in earth history. At least among the vertebrates, no single species is known to have colonized so many diverse environments in such a brief interval of time.[19] In a period of no more than 20,000 years—possibly much less—*Homo sapiens* occupied all of the areas in southern Eurasia colonized by *Homo erectus* 1.8–1.4 million years ago and also invaded Australia and much of northern Eurasia. The expansion of human settlement in cold regions at this time can be understood only within the context of the broader dispersal.

The dating of the dispersal is still a work in progress. Modern humans are clearly present in Southern Asia and Australia by 40,000 years ago and possibly much earlier, and they seem to be present in at least some parts of northern Eurasia by 45,000 years ago.[20] But the dispersal probably began at a significantly earlier date—in the range of 60,000–50,000 years ago. Much of the uncertainty about the chronology is tied to the limitations of radiocarbon dating. By 35,000 years, most radiocarbon in a dating sample has decayed. Samples from sites that are at least 40,000 years old are easily contaminated by younger radiocarbon and likely to yield false dates.[21] Alternative methods of dating, such as luminescence and electron spin resonance (ESR), are now being applied more widely, but the results are so far limited.[22] Other sources of chronological information, such as correlation with

Figure 5.2. Map showing dispersal of modern humans out of Africa and colonization of Eurasia between roughly 50,000 and 25,000 years ago.

buried soils, paleomagnetic stratigraphy, volcanic ash layers, and so forth, are sometimes helpful but also suffer from some imprecision and uncertainty.

To begin with, modern humans are present in the Levant as early as 120,000–90,000 years ago. A slightly archaic form of *Homo sapiens* is found at the sites of *Skhul* and *Qafzeh* in Israel. The dating of these specimens in 1987–1992 shocked many anthropologists, because these modern humans are significantly older than at least some of the Neanderthal remains in the Near East. The latter have yielded dates as young as 50,000–40,000 years old.[23] The late Neanderthals in the Levant seem to represent a southward intrusion from Europe during and after the Lower Pleniglacial (see chapter 4).[24]

The skeletal remains from Skhul and Qafzeh do not appear to be part of the dispersal event that brought modern humans to places like Australia and Siberia. They probably reflect a limited expansion of the African population into an adjoining area of the Levant—and apparently during a warm climate interval. Significantly, while these people exhibit the fully modern position of the vocal tract (indicated by the shape of the basicranium), they lack accompanying archaeological evidence for the use of symbols and innovative technologies that are associated with later modern humans.[25]

The true dispersal may have begun with an initial movement across the tropical zone of Eurasia. Dates as early as 62,000 years ago have been reported from modern human sites in Australia, although a revised date of 42,000 years announced recently from *Lake Mungo* suggests that the island continent might not have been invaded until later.[26] An indisputably modern human skull recovered some years ago in southern China at *Liujiang* (latitude 24° North) is now believed to date to at least 68,000 years. However, the provenance of the skull—and hence the new date—remains uncertain.[27] Although there is some supporting biomolecular data for early dispersal into tropical Eurasia,[28] it cannot be confirmed at present.

In the Near East, evidence for a technological shift toward production of stone tools on blades and some bone implements dates to as early as 47,000–45,000 years ago. Simple ornaments in the form of perforated shells also appear in this time range at sites such as *Ksar 'Akil* in Lebanon. But skeletal remains of modern humans are known so far only from younger contexts—the earliest seems to be the skull of a child recovered from Ksar 'Akil and dated to roughly 35,000 years ago.[29]

The first traces of modern human settlement in northern Eurasia may be from the East European Plain. Ongoing excavations at the *Kostënki* open-air sites, which are located about 250 miles (400 km) south of Moscow on the Don River (latitude 52° North) have yielded stone blade tools, bone implements, ornaments, and figurative art dating to about 45,000 years ago. Although only isolated human teeth have been recovered from the levels

Figure 5.3. Excavation at Kostënki on the Don River in Russia.

that yielded these artifacts, the latter were almost certainly made by modern humans.[30]

Artifacts produced by modern humans also appear in southern Siberia at about the same time or slightly later. At the open-air site of *Kara-Bom* in the Altai region (50° North), they are dated to 42,000 years ago. There are no skeletal remains from Kara-Bom, but the stone artifacts are typical of those found elsewhere with modern human fossils.[31] Isolated teeth assigned to the latter are known from two other south Siberian sites that are younger—probably 35,000–30,000 years old—and more complete skeletal material is unknown in the region until after the beginning of the Upper Pleniglacial.[32]

As modern humans spread into northern Eurasia, it appears that they were invading warmer areas like southwestern Europe only after first establishing themselves in colder and drier regions. Their earliest sites in Western Europe, such as *Willendorf* (Austria) and *El Castillo* (northern Spain), date to 40,000 years ago at the most.[33] There are few if any skeletal remains in these sites, but both modern human fossils and sites containing their artifacts become more common after 35,000 years ago.[34] A late arrival—perhaps as late as 32,000 years ago—now seems likely for modern humans in the warmest and southernmost parts of Eastern Europe (that is, the Crimean Peninsula and Northern Caucasus).[35]

This pattern of colonization is reversed from that of earlier humans, who initially occupied only the warmest parts of northern Eurasia (see chapter 3) and seems especially curious in light of the fact that the new arrivals

Box 3. Climate in the Northern Hemisphere: 60,000–20,000 Years Ago

The dispersal of modern humans into northern Eurasia took place during an interval of almost constantly alternating climates. Conditions were generally cooler than the present day, but apparently did not reach the extreme cold of the Lower Pleniglacial. The period between roughly 60,000 and 30,000 years ago—often termed the Middle Pleniglacial—was marked by a series of brief warm and cool oscillations. During the succeeding interval (Upper Pleniglacial), climates again descended into full glacial conditions, achieving maximum cold at 24,000–21,000 years ago.[a]

Many proxy records of past climate are available for the Middle Pleniglacial, but stable-isotope cores provide the best overall framework. The Greenland Ice-Core Project (GRIP) and Greenland Ice Sheet Project 2 (GISP2) have produced a particularly detailed record for this period based on fluctuations in the $^{18}O/^{16}O$ ratio. Reversing the pattern in the deep-sea cores, higher percentages of the lighter oxygen isotope (^{16}O) in the ice reflect glacier expansion and lower temperatures. Cold phases are also represented by dated ocean floor sediments (Heinrich layers) deposited from drifting icebergs.[b]

The ice cores revealed more than a dozen short warm intervals (Dansgaard-Oeschger events) between 60,000 and 30,000 years ago. Five of these intervals saw significant temperature increases and may be considered "interstadial" periods, although annual temperatures probably remained 9–11° F (5–6° C) below current levels. The study of pollen-spore samples, fossil insects, vertebrate remains, buried soils, and other sources of information about the environments reveal relatively mild conditions in parts of Europe and southern Siberia. Arboreal vegetation—including some deciduous broadleaf trees—was widespread in many areas, and cold-loving mammals were less common.

The warm intervals between 60,000 and 30,000 years ago were interspersed with equally brief cold oscillations. At least two of these correspond to Heinrich events. During the cool phases, temperatures fell as much as 9–14° F (5–8° C) below those of the interstadial intervals. In regions of northern Eurasia occupied by modern humans, arboreal vegetation declined and tundra beetles and arctic mammals such as polar fox increased in numbers.[c]

---

a. J. J. Lowe and M.J.C. Walker, *Reconstructing Quaternary Environments,* 2nd ed. (Edinburgh Gate: Longman Group, 1997), pp. 333–337.

b. Gerard Bond et al., "Evidence for Massive Discharges of Icebergs into the North Atlantic Ocean during the Last Glacial Period," *Nature* 360 (1992): 245–249; W. Dans-

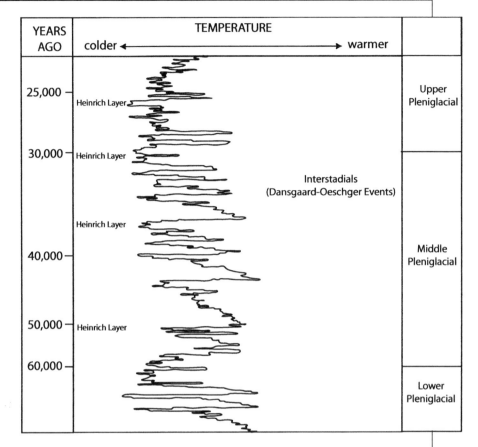

Figure B3. Temperature curve based on fluctuations in the ratio of oxygen isotopes from the GRIP Summit ice core.

---

gaard et al., "Evidence for General Instability of Past Climate from a 250-kyr Ice-Core Record," *Nature* 364 (1993): 218–220; P. M. Grootes et al., "Comparison of Oxygen Isotope Records from the GISP2 and GRIP Greenland Ice Cores," *Nature* 366 (1993): 552–554.

   c. A.A. Velichko et al., "Periglacial Landscapes of the East European Plain," in *Late Quaternary Environments of the Soviet Union*, ed. A.A. Velichko, pp. 94–118 (Minneapolis: University of Minnesota Press, 1984), pp. 113–114; Lowe and Walker, *Reconstructing Quaternary Environments*, pp. 336–337; Ted Goebel, "The Pleistocene Colonization of Siberia and Peopling of the Americas: An Ecological Approach," *Evolutionary Anthropology* 8 (1999): 208–209.

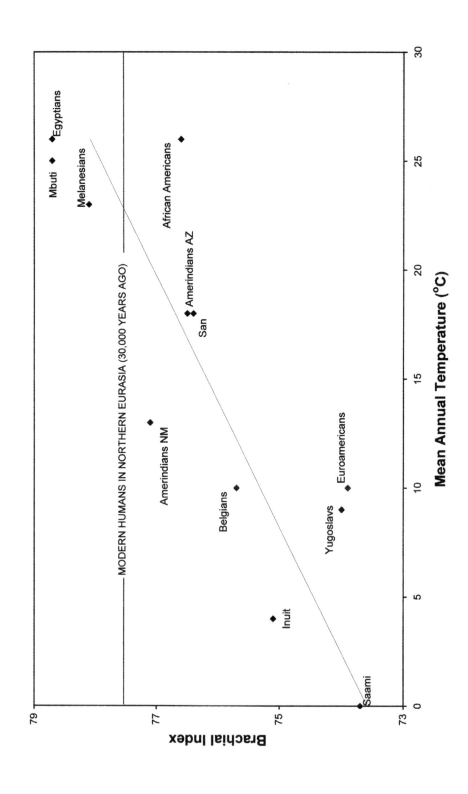

were also derived from southern latitudes. Although only isolated teeth and bones are known from the earliest modern human sites at latitudes above 40° North, complete or nearly complete skeletons have been recovered from European sites dating to 35,000–30,000 years ago. They include famous burials from caves at *Paviland* (England), *Cro-Magnon* (France), and *Grotte des Enfants* (Italy), as well as one from Kostënki (Russia). Measurement of their bones reveals that they possessed the body dimensions of tropical peoples.[36]

The dispersal of modern humans into northern Eurasia inverted the biogeographic "rules" of climate adaptation with respect to anatomy. Between 60,000 and 30,000 years ago, Europe and Siberia were experiencing an interval of oscillating climates—often termed the *Middle Pleniglacial*—between two major glacial advances. Even during warmer oscillations, however, climates were significantly cooler and drier than those of the present day. Mean January temperature for the East European Plain roughly 35,000 years ago (during the milder period preceding the Last Glacial Maximum) is estimated at $-5°$ F ($-20°$ C) and precipitation on the plain is thought to have declined by 25–35 percent.[37]

Because of their African physique, cold stress on the modern humans who first occupied Europe and Siberia must have been extreme—greater than the stress on more recent peoples in high latitudes—and it serves to emphasize the overwhelming importance of their innovative technology for coping with these environments. It also draws further attention to the curious pattern of initial settlement in colder and drier regions like the East European Plain.

In fact, it is now apparent that Middle Pleniglacial settlement of the East European Plain and Siberia extended as far as the Arctic Circle (66° North) and beyond. New dates on the site of *Mamontovaya Kurya* in northern Russia demonstrate at least seasonal occupation at this latitude. Today the area experiences a January mean of $-2°$ F ($-18°$ C) and climates at the time of occupation (roughly 40,000 years ago) were almost certainly colder. Even more remarkable is a recently reported site near the mouth of the Yana River at 71° North in northeastern Siberia, which dates to almost 30,000 years ago—a warm interval at the end of the Middle Pleniglacial.[38]

The belated appearance of modern humans in Western Europe is most probably a result of the Neanderthal presence in this region. This may explain the late arrival of modern humans in the southernmost parts of Eastern Europe. It will be recalled from the preceding chapter that the

Figure 5.4. Ratio of lower arm length to upper arm length (brachial index) in living peoples, showing the position of modern humans in northern Eurasia 30,000 years ago—close to living peoples of the tropical zone.

Neanderthals contracted their range during the Lower Pleniglacial and abandoned the coldest parts of northern Eurasia. Reoccupation of these areas during the succeeding Middle Pleniglacial seems to have been limited and sporadic. As modern humans moved north, they revealed a preference for places where the Neanderthals remained scarce or absent altogether. It was only later that they began to encroach on habitat still occupied by the native population and compete for the same resources. By 30,000 years ago, there were few if any Neanderthals left in Europe or elsewhere.

The Archaeology of Symbols

In 1879 Don Marcelino Sanz de Sautuola entered a cave on his property in northern Spain and discovered the spectacular cave paintings of *Altamira*. The polychrome images of bison painted on the ceiling became especially famous. Noting parallels with engraved images on bones recovered from Paleolithic sites in France, de Sautuola argued in a booklet published the following year that the Altamira paintings must be equally ancient. However, his thesis was attacked violently by other scholars—largely on the grounds that Paleolithic people would have lacked the artistic vision and skill to create such magnificent art.[39]

De Sautuola was eventually vindicated, of course, and by the early 1900s archaeologists recognized the paintings in Altamira and other Paleolithic caves as the work of the modern humans who lived during the last glacial period. Nevertheless many continued to assume that the most Paleolithic art was relatively late—generally younger than 20,000 years ago.[40] In recent years, the discovery and dating of new sculptures and cave paintings (including the direct radiocarbon dating of paintings with accelerator mass spectrometry) has shown that some of the most impressive works of Paleolithic art date to more than 30,000 years ago.[41] And a newly recovered ivory figurine head from the lowest levels at Kostënki suggests that art was present when modern humans first arrived in northern Eurasia.[42]

The visual art created by these people as they dispersed out of Africa remains the most striking evidence of their behavioral transformation. Art is so complex and uniquely human that its presence in the archaeological record—especially in such a highly sophisticated form—has convinced most anthropologists that the cognitive skills of these humans were fully modern. Although the anatomical evidence indicates that modern speech capabilities had evolved before 100,000 years ago, it is the appearance of art—fossilized symbolic structures—that confirms the presence of modern human language.[43]

Like spoken language, visual art provides a means of constructing and communicating concepts about the world with symbols. Although many

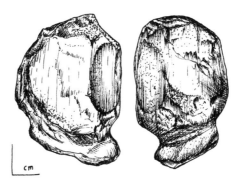

Figure 5.5. Head of a human figurine carved out of mammoth ivory, recovered from a layer at Kostënki 14 dating to more than 40,000 years ago.

works of art seem to mimic people, animals, and other objects, virtually all are symbolic constructs—similar to words and sentences—bearing only limited and arbitrary relation to their subjects. Even the most literal representations of animals in Paleolithic art are reduced in size (or collapsed into two dimensions in the case of paintings and engravings), and lacking the texture, anatomical detail, smell, sound, movement, and often the color of the depicted animal. They are interpretive models of selected and abstracted phenomena (often partly or wholly imagined) extracted from the continuum of experience.[44]

Equally impressive—but sometimes ignored by archaeologists—are the musical instruments in modern human sites dating to at least 35,000 years ago. Flutes in this time range have been recovered from the caves of *Isturitz* (southern France) and *Geissenklösterle* (Germany), and they are also present in many younger sites.[45] Some simple percussion instruments are reported from a Ukrainian site dating to roughly 25,000 years ago,[46] although these are problematic. Many musical instruments or parts of instruments may not have been preserved or recognized in the archaeological record.[47]

Music is another means of structuring and communicating—in this case with the use of "sonic symbols," or sounds in the audible spectrum. Studies of brain function reveal overlapping networks for speech and music in the neocortex, suggesting that the capacity for language and music evolved together.[48] Music is universal and—like spoken language—may assume very complex forms among the most "primitive" peoples.[49] And like spoken language and visual art, music is almost infinitely productive: sounds generated by instruments and voices are constantly rearranged to create new compositions.

Spoken language, visual art, and music are not the only means by which modern humans structure their surrounding world. They also classify and manipulate fundamental concepts about the perceived physical universe—

space, time, and visible light (that is, colors). By 30,000–25,000 years ago, the structured use of space is quite apparent from the arrangement of work areas (indicated by concentrations of stone and bone debris), hearths, dwellings, and sometimes storage pits on occupation floors at *Kulichivka* (Ukraine), *Mal'ta* (Siberia), and other sites.[50] The pattern contrasts sharply with that of earlier humans—even the Neanderthals—who organized their living space to a minimal degree.

The structuring of time is less obvious in the archaeological record, but simple calendars—fragments of bone, ivory, and stone bearing sequences of markings—may be present more than 35,000 years ago. Examples of possible lunar calendars are reported from *Border Cave* (South Africa) and the *Abri Blanchard* (France), as well as from other sites.[51] Although these are similar to "calendar sticks" found among tribal peoples in various parts of the world, some archaeologists have argued that they might represent something else.[52] Color classification—while universal among all recent and living peoples—is also hard to verify in a prehistoric context. However, the mixing and use of colored pigments in the cave paintings most probably reflect categorization of colors.[53]

Rivaling the paintings for splendor are the burials found in open-air and cave sites from Spain to Australia. Among the most spectacular examples are those at *Sungir'* in northern Russia (located northeast of Moscow at latitude 56° North) and dated to about 27,000–26,000 years ago. The graves include an older male and two adolescents—buried head to head—all of them laid to rest with a remarkable amount of funerary adornment and objects. Necklaces, bracelets, brooches, and thousands of ivory beads (apparently sewn into the clothing) covered the skeletons. Spears and other items had been placed alongside them in the grave, which was lined with charcoal and filled with red ochre.[54]

The burials at Sungir' and elsewhere are rich in various forms of symbolism. The ornaments on the deceased—also abundant in nonfunerary contexts—illustrate yet another aspect of categorization and structure. Modern humans classify each other socially and often dress themselves with clothing and adornment that signifies their position as well as their ethnic identity. The grave offerings and pigments sprinkled over the bodies imply rituals associated with death and burial, and these in turn undoubtedly reflect concepts and beliefs about afterlife and the spirit world.[55]

The totality of these rituals, concepts of space and time, musical compositions, art objects, and other symbols constitute what anthropologists long ago named "culture."[56] The presence of the material expressions of culture in the archaeological record suggests that its many nonmaterial forms—jokes, recipes, dances, chants, myths, and so forth—were present among modern humans at this time as well. Anthropologists also recognized long

Figure 5.6. Burial at the site of Sungir' in northern Russia.

ago that each culture comprises not a random assortment of symbols, but what Mary Douglas described as an "integrated whole."[57] Each culture offers a coherent worldview and a set of rules of etiquette and behavior that go with it.

As modern humans spread out of Africa, they carried with them this mass of structured symbols—culture. And it is apparent that culture changed and diversified rapidly as modern humans invaded a myriad of different habitats. The dispersal that took place roughly 50,000 years ago laid the basis for the linguistic and cultural diversity of the modern world (although much additional diversification followed toward the end of the last glacial period, as people spread into the Arctic and the Americas). Both the initial and the later dispersals offer a contrast to the comparatively limited genetic variability among living peoples given the extraordinary range of occupied habitats.[58]

Culture undoubtedly played an important role in the dispersal and the ability of modern humans to adapt so rapidly to so many different environments. One of its more obvious advantages is the capacity it provides for compiling a vast database—maintained as a collective memory bank among the group—on the plants, animals, and other features of a given habitat. Anthropologists studying tribal peoples in various parts of the world have found that the detail and precision of their plant and animal classification systems are often comparable to scientific taxonomy. Large amounts of information are also compiled on body parts, animal behavior, and the uses of plants. The classifications are incorporated into songs, myths, visual art, social categories, and other elements of the culture.[59]

Direct evidence for detailed plant and animal classification 50,000–30,000 years ago is admittedly rather thin. The many animal species and occasional plants depicted in paintings, engravings, and sculptures dating to the last glacial period are perhaps the only material expression of a system otherwise maintained through oral tradition. After 20,000 years ago, the taxonomic variety of the artistic images increased significantly, but the number of taxa represented remains only a fraction of the plant and animal categories found among recent and living peoples.[60] Classification systems among modern humans both before and after 20,000 years ago were probably far more elaborate than the record shows.

The ability to compile, organize, and share a mass of coded information on biota and landscape would have been an unprecedented advantage for the colonization of new lands. Many of the habitats invaded by modern humans after 50,000 years ago contained plant and animal taxa unknown in Africa, including unique species in Australia and the boreal flora and fauna of Siberia.[61] As among living peoples, the database could be modified to accommodate novel forms. It would have helped the response to temporal as well as spatial variations in environment. As local plants and animals changed during the climate oscillations of the last glacial period, modern humans could adjust their classificatory systems accordingly.

Culture probably offered another critical advantage for the colonization of new places—specifically colder and drier places. As noted in the preceding chapter, it is difficult to imagine modern human social life without the shared symbols that people use to establish and maintain their relations. Several archaeologists have argued that by facilitating the formation of large social networks, spoken language and other symbols were essential to the occupation of environments where food resources were thinly distributed—specifically Siberia and Australia. Tribal peoples in these regions traditionally relied on alliances among widely scattered families to share information, ensure adequate food supply, and maintain a marriage network.[62]

The archaeological record indicates a quantum leap in long-distance movements and contacts with the appearance and spread of modern humans. The movement of raw materials over great distances is one of the more striking changes that accompany the dispersal out of Africa. Between 40,000 and 30,000 years ago, modern humans were moving fossil shells from the Black Sea area across hundreds of miles to places like Kostënki on the central East European Plain.[63] Transport of shells and other materials over comparable distances—unprecedented in human evolution—is reported from other regions at this time.[64] Whether partly by trade between groups or not, the pattern is similar to that found among recent peoples in tundra and desert, who operate across large areas as part of alliance networks.[65] After 30,000 years ago, the distribution of certain art objects (most notably the famous "Venus" figurines of Central and Eastern Europe) indicate culture contacts over immense areas.[66]

It is probably a mistake, however, to explain every aspect of culture in terms of its adaptive value. In the early twentieth century, anthropologists like Bronislaw Malinowski and A. R. Radcliffe-Brown insisted that culture functioned to provide material and social benefits to individuals and groups. During the latter part of that century, cultural ecologists like Julian H. Steward described culture as a nongenetic (or "extrasomatic") means of adaptation. These views have been disputed—often bitterly—by other anthropologists, who argue that culture can be understood only on its own terms as a symbolic construction of the perceived and imagined universe.[67]

For every aspect of culture that reveals an adaptive function, there is another that seems to be a sheer waste of time and energy. Among all peoples, for example, the systems of plant and animal classification that contain so much useful information on food, medicine, materials, and dangers also include a mass of categories that are completely devoid of any practical value.[68] The same pattern is apparent in the poor correspondence between the animals depicted in the cave paintings of 35,000–20,000 years ago and those of economic importance in archaeological sites of the same age.[69] Here, as with other aspects of culture, the error lies in trying to understand parts isolated from the whole.

The capacity for culture—spoken language and all the other uses of symbols—must have evolved in modern humans because it conferred enormous advantages on those who possessed it. These advantages presumably outweighed the costs in time and energy. Furthermore, once established among a group, cultural behavior must have been reinforced by intense social pressure as among all living peoples. But most of the varied components of each culture—like the sound components of each language—are understandable only within the context of that culture.

The Birth of the Machine

Whatever the adaptive advantages of symbols, their appearance in the archaeological record coincides with an extraordinary transformation in human technological ability. As in the case of spoken language, modern humans seem to have left Africa with a technical skill and imagination fully commensurate with that of people today. The same sort of cognitive abilities that underlie the manipulation of words and other symbols found expression in the creation of complex tools, weapons, devices, and facilities.

Speculation about the relationship between language and tools has a long history. The American archaeologist James Deetz wondered some years ago if "words and artifacts are in fact different expressions of the same system," and Glynn Isaac believed that linguistic and technical abilities had evolved together.[70] Nevertheless, formal comparison of language structure and tool behavior reveals fundamental differences. The cognitive processes engaged in making tools have been compared to those of a prelinguistic child—not an adult speaking syntactical language.[71]

Because human technology probably has deep roots in the Miocene apes and a long evolutionary history that precedes any evidence for language in the archaeological record, it is not surprising to discover its early developmental character in young children. Moreover, while symbols are often incorporated into modern human technology (for example, a ship's name painted on its bow), the core components of a tool or device are *not* symbols, but natural materials that have been modified and combined to effect other changes on the environment.

But if modern human technology does not mirror the structure of grammar and syntax, it does seem to reflect the impact of what humans can do with language and other symbol systems. The technology of the people who dispersed out of Africa 60,000–40,000 years ago exhibits the structural *complexity* of the models or symbolic constructs (such as art objects, myths) that are found among all recent and living cultures.[72] Significantly, modern humans became the first to design mechanical tools and devices, which ultimately became the basis for industrial civilization.[73] And their technology also exhibits the *productivity* of language and symbolic constructs—the ability to create novel forms through continual recombination of elements. From this point onward, technology in the hands of modern humans reveals constant change and innovation.[74]

Many decades ago, the archaeological remains of modern humans in Eurasia were classified as *Upper Paleolithic* largely on the basis of artifacts recovered from deep occupation sequences in the caves of France.[75] As traditionally defined, Upper Paleolithic industries exhibit a general shift from the production of stone tools on flakes to those on blades. High percentages of certain tool types such as chisel-like burins, along with some implements of bone, antler, and ivory, are found. These industries also contain examples of art and ornament, which until recently were thought to be relatively simple in the early Upper Paleolithic.[76]

Most textbooks on human prehistory echo the classic definition of the Upper Paleolithic and fail to convey the revolutionary nature of the changes in the archaeological record that inaugurate this period. This reflects in part the traditional perspective of West European archaeologists viewing the Paleolithic as a sequence of progressive evolutionary stages. It also

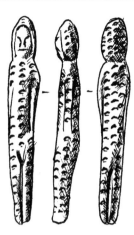

Figure 5.7. Carved figurine from the site of Buret' in Siberia, which probably dates to about 25,000 years ago, showing use of tailored fur clothing.

reflects the fact that much of the evidence for complex technology is based on inference and nonarchaeological sources of information.[77]

The ability to create innovative and complex technology was instrumental in the dispersal of modern humans out of Africa between 60,000 and 40,000 years ago. Their appearance in Australia was achieved through the construction of watercraft required to navigate a minimum 55 miles (90 km) of open water, which had presented an absolute barrier to earlier human settlement.[78] Complex technology was especially critical to the occupation of higher latitudes in Eurasia, and probably the most important variable in the expansion into cold environments not previously inhabited by the Neanderthals.

Modern humans developed new technology related to cold protection that was essential for survival in northern Eurasia. Production of tailored fur clothing is indicated by the appearance of eyed needles of bone and ivory on the East European Plain as early as 35,000 years ago, and perhaps in southern Siberia at roughly the same time.[79] Needles are completely absent from Neanderthal sites. Carved figurines from the slightly younger Siberian site of *Buret'* depict people clothed in fur suits—complete with hoods.[80] Upper Paleolithic scraping tools generally exhibit heavier microscopic polish than those from Neanderthal sites, reflecting intense hide preparation.[81]

The first convincing traces of artificial shelters are also found in Upper Paleolithic sites of northern Eurasia. Here again the evidence is found primarily on the East European Plain and in southern Siberia, where natural shelters are rare. Traces of small artificial shelters with interior hearths—presumably tents composed of hides attached to a wooden frame—have been reported from several occupation floors at least 30,000 years old (for example, Kulichivka in Ukraine). They are more common in younger sites.[82]

Modern humans also employed their technological imagination to expand the range of exploited animals to smaller mammals, birds, and fish—species that seem to have been inaccessible to their predecessors. Most of this evidence is based on indirect sources. The remains of small mammals such as fox and hare are found in large quantities in East European sites dating to at least 40,000 years ago, and these probably reflect the use of snares or traps. Hares might have been caught in nets, production of which is indicated by clay impressions at several younger sites (such as *Zaraisk* in Russia). Although the hares were probably consumed as food, the foxes—often found as nearly complete skeletons—were used primarily for their pelts.[83]

Stable-isotope analyses of human bone indicates heavy consumption of freshwater aquatic foods in the form of waterfowl and/or fish in sites dating to at least 30,000 years ago.[84] Although few archaeological traces of the technology required to harvest large numbers of birds and fish are known to date (for example, a hook fragment from *Mezin* in Ukraine),[85] it may have included throwing-board darts, nets with sinkers, and various other novel devices. Modern humans were using technology to broaden their ecological niche in northern Eurasia far beyond the boundaries of the Neanderthal niche. But like the Neanderthals, they were also hunting large mammals such as horse and reindeer, which apparently provided much of their high protein/fat diet.

Perhaps the defining difference between the technology of modern humans and that of all their predecessors was the presence of mechanical devices (implements or facilities with moving parts). The appearance of mechanical tools and weapons in Upper Paleolithic sites was noted long ago by V. Gordon Childe.[86] The conceptual basis of mechanical technology may lie in the ability to create complex models with syntactical language and other symbols, and thus beyond the grasp of the premodern human brain.

Convincing archaeological evidence for mechanical technology is so far confined to after 20,000 years ago—following the reoccupation of areas abandoned during the cold peak of the Upper Pleniglacial (see chapter 6). However, at least some of the technology developed between 45,000 and 24,000 years ago may have included mechanical devices. Tailored fur clothing might have been equipped with drawstrings (found in Inuit clothing), while some traps and snares are designed with moving parts. Fire making and fowling and fishing technology also could have included some mechanical items.[87] Future discoveries may alter the current picture.

Despite its enormous practical value—and central role in the settlement of higher latitudes—the ingenious technology of modern humans can be fully understood only within the cultural context of the people who created it. For example, one of the most impressive technical achievements of this period was the development of fired ceramics. By roughly 30,000 years ago, people in Central Europe were building and heating kilns up to 800° C to produce fired clay objects. The latter were figurines—not pots or other items with a recognizable economic function—and there is some evidence that they were intentionally exploded as part of a ritual process.[88]

The Last Glacial Maximum

After roughly 28,000–27,000 years ago climates began to cool again in the northern hemisphere. Winter temperatures eventually surpassed those of the Lower Pleniglacial, reaching an estimated January mean of −22° F (−30° C) on the central East European Plain. Glaciers expanded to cover much of northwestern Europe during the cold peak between 24,000 and 21,000 years ago. Fierce winds generated along the glacier margins blew large quantities of silt (periglacial *loess*) out over the plains of northern Eurasia. Trees disappeared from many areas where conditions had become too cold and dry.[89]

By 28,000 years ago, the Neanderthals were gone and modern humans occupied much of Eurasia. As climates cooled during the early phase of the Last Glacial Maximum, the moderns' sites actually increased in size and numbers. Between 28,000 and 24,000 years ago, people apparently thrived in Europe and southern Siberia, sometimes occupying large and complex living sites and creating spectacular art. But as the maximum cold peak approached 24,000 years ago, people abandoned these regions for several thousand years, despite the fact that conditions at this time were probably no worse than those endured by some modern Arctic peoples.

Archaeologists often designate this interval as the "middle Upper Paleolithic" because the sites are easily distinguished from those of other periods. Many of the European sites that date to between 28,000 and 24,000 years ago are assigned to the *Gravettian* industry. They contain rather characteristic shouldered points of stone and the famous "Venus" figurines of unclad females. A broadly similar set of remains are found in Siberia at this time, although the shouldered points are absent and the style of the female figurines is different.[90]

In the coldest and driest areas occupied by the Gravettians—the central plain of Eastern Europe—it is possible to see the beginnings of an adaptation to arctic conditions, although plant and animal productivity was probably much higher than modern tundra because of the comparatively low latitude (combined with the nutrient-rich loess).[91] The parallels between the Gravettian (as well as other European Upper Paleolithic industries) and the later peoples of the Arctic were noted as early as the 1860s.[92] Like the Inuit and other such peoples, the Gravettians inhabited a land without trees and were forced to use substitutes for fuel and raw material.

They created a varied and complex technology of hide, bone, antler, and ivory, and many of their artifacts are remarkably similar to those of the Inuit. They fashioned mattocks from ivory and used them to dig large storage pits, kept their sewing needles in small cases of bone, manufactured portable lamps fueled with animal fat, and constructed a semi-subterranean house—probably for winter use—at *Gagarino* on the Don River (Russia). Microwear analysis of their ubiquitous shouldered points indicates that their concave edges were actually used for cutting hide like an Inuit woman's knife *(ulu)*.[93]

Lacking wood, the Gravettians gathered large quantities of mammal bone for their hearths. In many places, they may have been able to scavenge natural concentrations of bone that had accumulated at stream confluences and ravine mouths. The large pits or caches found in most of their sites are often thought to have been used primarily for meat storage, but may have also functioned as "ice cellars" in warmer months for keeping bone fuel fresh and flammable.[94]

The heavy consumption of bone in the hearths complicates the analysis of the Gravettians' diet—they seem to have burned up most of their food debris. Nevertheless, various lines of evidence suggest that reindeer, horse, and sometimes mammoth were being hunted. Stable-isotope analysis of human bone indicates exploitation of fish and/or waterfowl, and this might explain some of the more ambiguous items in their artifact assemblages, which resemble net-sinkers, bolas, and components of other devices used by the Inuit for fishing and fowling. Traps and/or snares were employed to harvest wolves, foxes, and hares, which are abundant in Gravettian sites.[95]

Although evidence for untended facilities (traps and snares) is present prior to the Last Glacial Maximum, storage pits are documented for the first time in the Gravettian. Both are stratagems used by recent northern peoples to cope with the spatial and temporal distribution of resources in cold environments. Untended facilities not only permit harvesting of animals that are difficult to catch and kill, but also save considerable time and energy in habitats where resources are scattered and unpredictable. Storage of food—and perhaps conservation of fresh bone for fuel—helps counter the effects of pronounced seasonal variations in these habitats.[96]

Many of the Gravettian sites on the central East European Plain are enormous and indicate social gatherings of unprecedented size. Sites at Kostënki, as well as other Russian localities such as *Avdeevo* and Zaraisk, reveal complex arrangements of hearths and pits covering large areas. These sites seem to have been occupied—perhaps chiefly during the warm

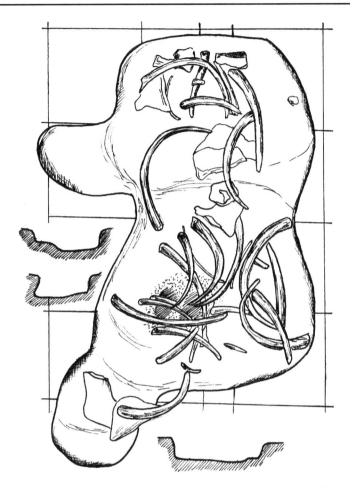

Figure 5.8. Large storage pit containing bones at the Gravettian site of Kostënki 1 in Russia.

season—by groups of fifty people or more (based on the number of hearths). They may correspond to seasonal concentrations of specific resources such as fish and/or waterfowl. Here again the pattern is common among the Inuit and other later arctic peoples, and reflects a strategy of coping with resources that are generally scarce but at times locally and temporarily abundant.[97]

Periodic aggregations of families otherwise scattered across large expanses of territory provide a means of sustaining a social network in such environments. At sites like Avdeevo, Gravettians probably assembled to reinforce kinship and economic ties through public ritual and other expressions of a shared ethnic identity. The geographic distribution of some of

Figure 5.9. Feature complex comprising hearths and pits at the Gravettian site of Zaraisk in northern Russia.

their material items—most notably the classic "Venus" figurines from the Danube Basin to the Don River—suggests that the social networks of the Gravettians were also of unprecedented size.[98]

A somewhat different way of life was pursued by the people who inhabited southern Siberia during the early Last Glacial Maximum. Despite the severe continental regime, estimated winter temperatures were probably no lower than those at comparable latitudes on the central East European Plain. Pine trees were present in some areas, providing wood for fuel and material.[99] Mal'ta and Buret' were occupied at this time, and contain traces of dwellings and storage pits. Mammoth, reindeer, and other large mammals were hunted, along with various birds.[100]

Archaeologists began to suspect a settlement hiatus during the peak of the Last Glacial Maximum in some parts of northern Eurasia years ago, but supporting evidence was slow to accumulate.[101] With hundreds of radiocarbon dates now available for Eastern Europe and Siberia, a severe depression between roughly 24,000 and 20,000 years ago has become apparent.[102] Much of the East European Plain and Siberia seems to have been abandoned at this time. The northern areas of Western and Central Europe were also vacated. People continued to dwell in southwestern Europe, where climates were predictably milder, but their lives were not unaffected. During the cold peak, sites are dominated by the *Solutrean* industry with its distinctive foliate points, and eyed needles appeared for the first time in Western Europe.[103]

The maximum cold of 24,000–20,000 years ago was an environmental disaster for modern humans in Europe and Siberia, and it brought to a close the epoch that had begun with the dispersal out of Africa. The people who reoccupied the coldest parts of Eurasia after 20,000 years ago were physically and culturally distinct from the earlier population, and they inaugurated a new period of prehistory and expansion into higher latitudes.

The reason why modern humans were forced out of much of northern Eurasia remains something of a puzzle. Adequate food resources in the form of large mammals were probably still available in many of the abandoned areas. This is suggested by the mass of radiocarbon dates for the cold maximum on mammoth, horse, bison, and other large mammals from the Bering Land Bridge, where climates were equally cold and dry.[104]

The explanation may lie in modern humans' anatomy, which still retained most of the tropical character of their African ancestors as late as 24,000 years ago (discussed earlier).[105] Measurable skeletal materials have yet to be found in Siberia, but assuming that the anatomical features of the Siberians were similar to those of the Europeans, all of these populations would have been susceptible to high rates of cold injury in very cold and dry conditions.[106] Medical data indicate that living people with comparable body dimensions—even when protected by insulated clothing—are subject to particularly high incidence of frostbite at low temperatures.[107] This could explain why modern humans were unable to tolerate mean January temperatures of $-22°$ F ($-30°$ C) prior to 20,000 years ago, when later arctic people with similar technology (for example, the Yukaghir) were able to survive under even harsher climatic conditions.[108]

CHAPTER 6

Into the Arctic

Books about human prehistory usually draw a line between the end of the Pleistocene or Ice Age—dated to about 12,000 years ago—and the epoch that followed. By this time climates were approaching those of the present day, and the cold-loving biota that had thrived at midlatitudes during the Upper Pleniglacial were retreating northward or disappearing altogether. The massive glaciers that had covered most of northwestern Europe and northern North America were melting, although the process of deglaciation continued for another 5,000 years. Archaeologists link these environmental changes to the end of the Paleolithic and the advent of global culture change—leading in some areas to agriculture and civilization.[1]

In fact, the period between the end of the Last Glacial Maximum cold peak and the end of deglaciation (roughly 20,000–7,000 years ago) was one of continuous climate and culture change. The boundary recognized at 12,000 years ago is somewhat arbitrary and its significance for human prehistory is overstated.[2] For the purposes of this book, it makes more sense to consider the period as a single unit.

Major developments took place during this period in terms of human settlement in the higher latitudes. After 20,000 years ago, people reoccupied the areas in Europe and Siberia abandoned during the cold peak.[3] And with the end of the Upper Pleniglacial, they established a sustained presence above the Arctic Circle for the first time.[4] This was possible, it seems, only through the conjunction of modern human technological abilities and the effects of climate warming. In some regions, anatomical adaptations to the cold—which emerged among modern humans only after 20,000 years ago—may have been an important factor as well.[5] As climates continued to warm, midlatitude Eurasia ceased to be a northern frontier of hominid settlement.

The most dramatic consequence of the occupation of higher latitudes after 20,000 years ago was the peopling of the New World. As modern humans expanded their range above 60° North in Siberia, they moved into the western portion of *Beringia*—the subcontinent joining Northeast Asia to Alaska during intervals of lowered sea level. People were probably living in eastern Beringia by 15,000 years ago—if not earlier—and they appear in other parts of North and South America shortly thereafter.[6] Occupation of the central and eastern Arctic, however, was delayed for thousands of years by the slow pace of deglaciation in Canada.[7]

Our knowledge of this critical period in prehistory is uneven. The peoples who occupied Western Europe and recolonized the colder regions of mid-latitude Eurasia during 20,000–15,000 years ago are well known. From Spain to Japan, this was the high watermark of the Upper Paleolithic in terms of the size and number of archaeological sites. By contrast, information about the early settlement of the Eurasian Arctic and Beringia after 15,000 years ago is limited. Much of the archaeological record is confined to small scatters of stone artifacts that tell us almost nothing about the way of life of these people.

Western Europe

The famous rock shelter of *Laugerie-Haute* overlooks the Vézère River in southwest France. More than 15 ft (> 5 m) of deposits here yielded a deep sequence of occupation layers initially examined in 1863 by Edouard Lartet, who used the site to help develop the original chronology for the Upper Paleolithic. Laugerie-Haute reveals a record of continuous occupation through the coldest phase of the Last Glacial Maximum and the milder period that followed. Layers dating to the cold peak (24,000–21,000 years ago) contain artifacts of the distinctive Solutrean culture (mentioned briefly in the previous chapter), while the overlying layers contain remains of the *Magdalenian* industry.[8]

The maximum cold phase brought subarctic climates to southwest France that were only slightly warmer than those of the East European Plain during Gravettian times. By some estimates, January temperatures ranged between −22° F (−30° C) and 5° F (−15° C), while annual precipitation declined to 16 inches (400 mm). Trees became scarce (confined to Scotch pine during the coldest phase) and steppic plants appeared. Reindeer dominate the large mammal remains at Laugerie-Haute and other French rock shelters inhabited at this time.[9]

A burst of technological innovation accompanies the maximum cold and its aftermath. Although mechanical devices may have been invented prior to the Last Glacial Maximum (see chapter 5), they are demonstrably present only in the waning centuries of the cold peak. The earliest known spear-thrower was recovered from a Solutrean layer dating to roughly 21,000 years ago at *Combe-Saunière I*. Spear-throwers are more common in the Magdalenian and typically decorated with engraved and carved animal images.[10] The bow and arrow also may have been developed in the Solutrean—possibly indicated by the spread of small stone points and backed bladelets at this time—but is firmly documented only in the later Magdalenian (roughly 14,000 years ago).[11] Both devices presumably increased hunting efficiency.

Box 4. From the Last Glacial Maximum to the Atlantic Period: 20,000–7,000 Years Ago

The period during which humans reoccupied areas abandoned at the time of the Last Glacial Maximum peak cold and subsequently expanded into the Arctic was characterized by warming climates in the northern hemisphere. From the extreme low temperatures of the Last Glacial Maximum—when mean annual values fell as much as 30° F (16° C) below current levels—climates gradually approached the warmest period in the last 100,000 years (Atlantic period). During this process, only one brief but intense cold oscillation occurred (*Younger Dryas* event) roughly 13,000–12,000 years ago.

As in the case of the preceding period (see box 3 in chapter 5), a detailed record of climate change is available for this interval. Moreover, the past 20,000 years fall well within the effective range of radiocarbon dating, which can be calibrated to calendar years for an accurate and precise chronology. A high-resolution climate proxy record for the northern hemisphere is provided by the stable-isotope measurements from the Greenland ice cores (GRIP and GISP2). The oxygen-isotope curves can be correlated with estimates of *sea surface temperatures* (SST) in the North Atlantic on the basis of analysis of diatoms and foraminifera recovered from ocean floor sediments.[a] Numerous dated climate proxy records, including pollen-spore assemblages, fossil beetles, molluscs, and others allow terrestrial correlation of the ice core and marine records.[b]

Temperatures on sea and land remained low during the first four to five thousand years following the Last Glacial Maximum cold peak (although a brief minor warm oscillation took place about 18,000 years ago). The first significant warming is not recorded until after 16,000 years ago. Fossil beetle assemblages in northwest Europe indicate rapid temperature rise—estimated as much as 13° F (7° C) in a century—between 15,000 and 13,000 years ago *(Lateglacial Interstadial)*. This was followed by the Younger Dryas cold event (roughly 13,000–11,600 years ago), when annual temperatures declined by as much as 9° F (5° C).[c] A more or less steady warming trend ensued after the end of the Pleistocene, and temperatures in the northern hemisphere reached their highest levels during the Atlantic period (7,000–4,000 years ago) since the Last Interglacial climatic optimum (OIS 5e).

Some of the new technology was almost certainly tied to colder climates. Though already present on the East European Plain for more than ten thousand years, eyed needles appeared for the first time in Western Europe during the cold peak (the Solutrean).[12] The Magdalenians made increasing use of bone and antler technology (although the diversity of tools and

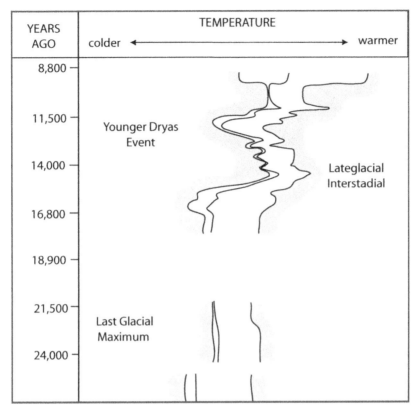

Figure B4. Estimated maximum summer temperatures based on dated beetle remains from the British Isles.

---

a. N. Koc Karpuz and E. Jansen, "A High-Resolution Diatom Record of the Last Deglaciation from the SE Norwegian Sea: Documentation of Rapid Climatic Changes," *Paleooceanography* 7 (1992): 499–520; Gerard Bond et al., "Correlations between Climate Records from North Atlantic Sediments and Greenland Ice," *Nature* 365 (1993): 143–147.

b. J. J. Lowe and M.J.C. Walker, *Reconstructing Quaternary Environments*, 2nd ed. (Edinburgh Gate: Longman Group, 1997), pp. 342–355.

c. Scott A. Elias, *Quaternary Insects and Their Environments* (Washington, DC: Smithsonian Institution Press, 1994), pp. 79–87.

---

weapons still seems low compared with the earlier Gravettians in Eastern Europe). Harpoons with either single or double rows of barbs are especially characteristic. They were carved with burins, the chisel-like tools that became common among their stone artifacts.[13]

The cold peak and its aftermath also witnessed a broadening of the range

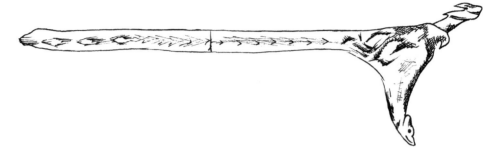

Figure 6.1. Spear-thrower from a Magdalenian site in France.

of food resources. During Solutrean times, reindeer were the primary focus of hunting in southwest France. In northern Spain, a variety of large mammals were hunted—including some mountain species—along with some birds and fish. Shellfish were collected at coastal sites, possibly during the winter months. The Magdalenians further expanded the diversity of animal foods, consuming larger numbers of birds and fish (both freshwater and marine) and gathering shellfish in large quantities.[14]

The overwhelming majority of Upper Paleolithic art in Western Europe dates to the Magdalenian period, which has been referred to as a creative "explosion." In addition to numerous cave paintings and engravings in southwest France and northern Spain, sculptures, engraved pieces, and other examples of portable art are common. Animals are a major theme (and they include not only large mammals but also fish, birds, and even invertebrates), although humans are sometimes depicted, as well as various geometric and abstract designs.[15]

It is widely believed that southwestern Europe became a crowded place during the coldest phase of the Last Glacial Maximum, largely as a result of the influx of people from northern areas abandoned after 24,000 years ago. In addition to the increased number and size of sites, there is evidence of nutritional stress (for example, enamel hypoplasias on human teeth). Some have argued that the changes in the archaeological record during this period—such as the more efficient hunting technology and broader range of food sources—were largely a response to the crowding and population pressure.[16]

Many of the observed changes in technology, economy, and organization in Western Europe after 24,000 years ago mirror earlier developments in Eastern Europe and Siberia, and they probably reflect a similar response to colder climates. Because of the moderating influence of the oceanic effect (see chapter 3), the cooling trend that began 28,000 years ago did not have a severe impact on Western Europe until the maximum cold phase. In many respects, the Solutrean and Magdalenian seem to parallel the Gravettian "arctic adaptation" in Eastern Europe of 28,000–24,000 years ago, described in the preceding chapter.

Climates ameliorated rapidly after 16,000–15,000 years ago and new changes began to sweep across Western Europe. Sites reappeared in northern France, Belgium, northwest Germany, and southern Britain between 15,500 and 14,000 years ago. Although many of them are considered late Magdalenian, other industries emerged in the northern areas containing distinctive curved-back and tanged stone points. An emphasis on reindeer hunting persisted through a cold oscillation during 13,000–11,600 years ago (Younger Dryas event). Deglaciation resumed after this cold event, and reindeer were quickly replaced by red deer, boar, and other forest fauna in northwest Europe.[17]

As the massive Fennoscandian ice sheet continued to shrink, plants, animals, and people expanded into freshly deglaciated areas of southern Scandinavia. Between 12,000 and 10,000 years ago, sites were occupied along the coasts of western Norway and southern Sweden up to latitude 65° North. These sites—which are assigned to the *Fosna-Hensbacka* complex—contain tanged points and other artifacts similar to those of earlier sites in northwest Germany and adjoining areas. They are found on islands and are thought to reflect a focus on marine resources, but little is known about the diet and economy of their occupants.[18]

The initial colonization of the arctic zone seems to have taken place at some point after 10,000 years ago. Sites assigned to the mysterious *Komsa* culture and dated to more than 7,000 years ago are found along the coast of Norway's Finnmark county above 70° North and further east on the Kola Peninsula. The Komsa sites are largely confined to surface scatters of stone artifacts, which include tanged points, burins, scrapers, and adzes. The stone—quartz, quartzite, and dolomite—is of local origin and poor quality. Several sites in the Varanger Fjord area also revealed traces of structures. These include shallow circular pithouses with diameters of 9–15 ft (3–5 m) and remains of larger rectangular dwellings occupying areas of 45–60 ft^2 (15–20 m^2).[19]

In many respects, the West European arctic zone retains a subarctic character similar to southern Alaska. The Atlantic Gulf Stream brings relatively warm and moist climates to coastal areas of Norway above the Arctic Circle, often creating storms and heavy fog banks. The ocean does not freeze during the winter, and coniferous forest extends all the way to the Arctic Ocean shore. Only the two months of winter darkness serve as a reminder of the high latitude.[20]

Despite a subarctic climate setting, the cultures that later developed in northern Norway had much in common with those of the Siberian and North American Arctic. In fact, it was a Norwegian archaeologist—Gutorm

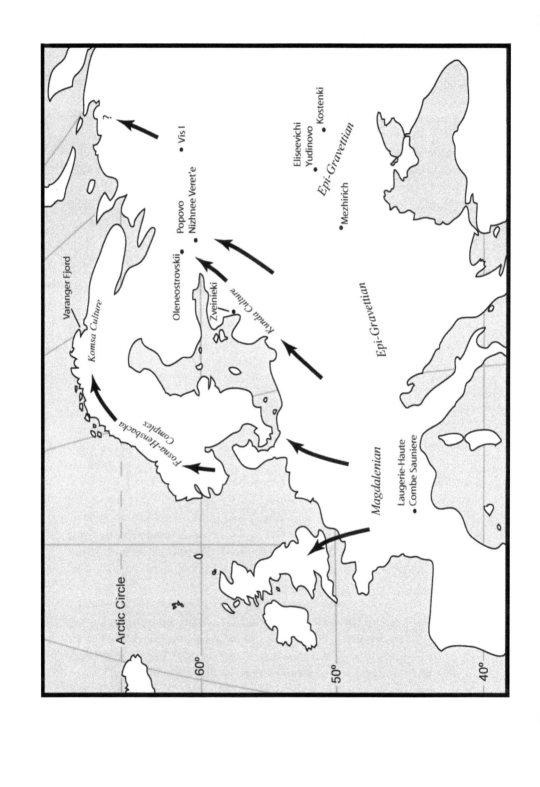

Gjessing—who first proposed the concept of a circumpolar culture. This notion was inspired as much by ethnographic parallels between the Saami people of Norway and the Inuit as by similarities among prehistoric artifact types. The Saami also evolved many of the same anatomical cold adaptations found among other high-latitude peoples.[21]

The sparse array of material remains recovered from the earliest sites in subarctic and arctic Europe reflects the bias of the archaeological record. And it was Gjessing—drawing on his knowledge of cold environments and the demands they make on people—who recognized that the Komsa sites preserve only a tiny fraction of the material culture created by their occupants. Their sites probably included skin boats, mechanical hunting weapons, portable lamps, insulated winter houses, rock art, and much more.[22] The antiquity of the sites and the conditions of their burial—or lack of burial—have erased most of the detail of the initial settlement of the West European Arctic.

Houses of Mammoth Bone

Though not covered by ice, the central plain of Eastern Europe remained largely or wholly uninhabited for several thousand years during the peak of the Last Glacial Maximum. Sites began to reappear about 20,000 years ago—in conjunction with a brief warm oscillation—although the scarcity of wood charcoal in them makes it difficult to obtain accurate radiocarbon dates. Nevertheless, a growing number of dates on bone and burned bone, combined with the stratigraphic position of the remains, indicate reoccupation of the plain up to latitude 53° North between 20,000 and 14,000 years ago.[23]

Despite the amelioration of climates, the East European Plain remained an extremely cold and dry environment. Periglacial loess continued to accumulate, and local plant and animal life retained a significant arctic component (including polar fox and musk ox). Temperature and precipitation were similar to those of Gravettian times (that is, prior to the cold maximum), and trees were absent across much of the central plain.[24] But as during the earlier period, plant and animal biomass was probably higher than that of arctic landscapes today.

The people who reoccupied the East European Plain after 20,000 years ago were different both physically and culturally from their Gravettian predecessors. The archaeological remains from this time range—not only

Figure 6.2. Map of northern Europe showing areas reoccupied following the cold peak of the Last Glacial Maximum (roughly 20,000 years ago) and initial colonization of the subarctic and arctic zones.

in Eastern Europe but also in Central Europe—are usually lumped under the term *Epi-Gravettian*. This is a broad term that undoubtedly subsumes a great deal of linguistic and cultural variability over a vast area and extended period of time.[25] The Epi-Gravettian sites of the central East European Plain exhibit some distinct patterns of their own, many of which reflect adaptation to the arctic conditions that continued to prevail in this part of Europe.

For the first time, modern humans exhibit at least some of the anatomical adaptations to cold climate that became characteristic of later peoples of the circumpolar zone. Human skeletal remains in European sites that postdate the cold maximum reveal shorter limbs than their predecessors. Although most of these remains are from sites in Western Europe, the same pattern is evident further east. A partial skeleton recovered from one of the younger localities at Kostënki (Russia) reveals a lower-limb length slightly below the West European mean.[26]

It is by no means clear that the changes in anatomy were critical to the reoccupation of the central East European Plain or other areas that had been abandoned during the cold maximum. After all, the Gravettians had endured similar climates prior to 24,000 years ago with their tropical physique. The reduction in limb length may simply have been a consequence of the catastrophic effects of the cold peak and—while obviously an advantage to any people living in arctic climates—not necessarily critical to maintaining a population in such climates. For modern humans, technology was the critical factor.

The most striking feature of the Epi-Gravettian sites is the houses of mammoth bone constructed in the central East European Plain. At least twenty of these remarkable former structures have been uncovered—sometimes in groups of three or four—in Ukraine and Russia. The remains of mammoth-bone houses are found in sites that lack wood charcoal, and they were probably a response to an absence of wood for shelter construction.[27] In a similar fashion, the Inuit would later use whalebone (along with driftwood) to build houses on arctic shores.[28]

Two sites investigated in recent years that have each yielded traces of four separate mammoth-bone houses are *Mezhirich* (Ukraine) and *Yudinovo* (Russia). The house remains are circular or oval in shape, and most of them range between 12 and 18 ft (4–6 m) in diameter. Limb bones were often used for the walls, although other skeletal parts—especially crania and jawbones—were incorporated into the structure. Lighter and flatter bones—shoulder blades and pelves—seem to have been used for the roof. Presumably mammal hides were attached to the bone framework for insulation. Each house typically contains one or more former hearths filled with burned bone and ash.[29]

Large pits filled with bone are found around the outside perimeter of the

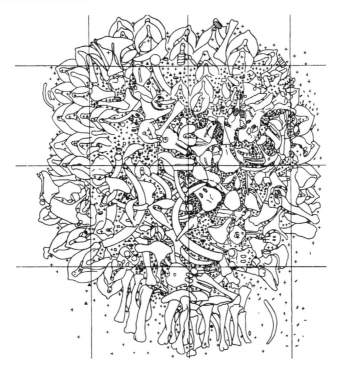

Figure 6.3. Remains of a mammoth-bone house at the Epi-Gravettian site of Mezhirich in Ukraine.

mammoth-bone houses. Many of the houses are surrounded with as many as four pits. They are similar to the pits in the earlier Gravettian sites (see chapter 5) and apparently were also used for cold storage of food and/or bone fuel. As in the case of the earlier sites, much of the large mammal bone in the Epi-Gravettian occupations may have been gathered from natural concentrations of animal remains that probably accumulated at ravine mouths—features that usually lie nearby these sites.[30]

The groups of houses found at sites like Mezhirich and Yudinovo are reminiscent of the layout of residential base camps among recent foraging peoples. If all of the houses at these sites were inhabited simultaneously—and there is supporting evidence that at least some of them were—they probably represent similar camps occupied by related families for several weeks or months.[31] This suggests that the Epi-Gravettians may have been able to establish small villages for extended periods. At the same time, raw materials from distant sources in these sites (such as amber, fossil seashells) suggest the operation of networks over enormous areas.[32]

For recent hunters and fishers, village settlements are tied to temporary abundances of resources and long-term storage of food. The pattern is common in arctic and subarctic environments, especially where rich concentrations of aquatic resources are found along rivers or ocean shores. However, it remains unclear what sort of temporary food surpluses (for example, reindeer herd, fish run) were available to the Epi-Gravettians. Like their predecessors on the East European Plain, they probably burned most bone debris of their large mammal prey, and stable-isotope data on their diet are not available. And the bone in the storage pits may not represent the pits' original contents.[33]

Aside from the mammoth-bone houses, much of the technology developed by the Epi-Gravettians was similar to that of their Gravettian predecessors. The wave of innovation that is thought to have occurred in Western Europe at this time is less evident on the East European Plain. At least part of the contrast—for example, the late invention or adoption of eyed needles—reflects the delayed impact of Last Glacial Maximum climates on Western Europe.

A wide variety of implements made from bone, antler, and mammoth ivory are found in the Epi-Gravettian sites, and the burins used to carve these materials often compose more than half of the stone tools. The spearthrowers made by the Epi-Gravettians' Magdalenian contemporaries in Western Europe are absent, but projectiles of some sort may be indicated by slotted bone points and large numbers of small stone bladelets. Enormous quantities of fur-bearing mammal bone (including many thousands of polar fox bones at *Eliseevichi* in Russia) suggest continued use of trapping equipment, and some possible trap components are reported from Mezhirich.[34]

Eliseevichi also contained two skulls identified as domesticated dog—apparently the oldest-known specimens on Earth (roughly 18,000 years old). Like all domesticated organisms, dogs are another example of modern human ability to manipulate the environment in complex ways. They may represent the earliest known form of biotechnology, and their appearance in these Epi-Gravettian villages of the East European Plain may be tied to the first long-term occupations in a cold-climate setting. The existence of semipermanent camps may have been a prerequisite for the domestication process. Dogs probably played an important role in the economies of people in cold environments from this point onward, although they are not always visible in the archaeological record.[35]

One of the most intriguing differences between the earlier and later inhabitants of the central East European Plain lies in their visual art. For reasons that will perhaps remain forever obscure, the Epi-Gravettians produced a more abstract set of sculptures and engravings. They made heavy use of geometric patterns, such as diagonal cross-hatching.[36] How these abstractions reflected their worldview and how that worldview was tied to their economy and organization are of course unknown. However, all of them would seem to have differed from the earlier Gravettian pattern, despite the similar environmental setting.

Boreal Forest Man

The Epi-Gravettian economy and worldview must have experienced drastic changes about 12,000 years ago, as climate warming and deglaciation transformed the East European landscape. The changing climate may have been somewhat slower to effect the latter. During the period that the Magdalenians were reoccupying northwest Europe (15,500–14,000 years ago), there is little evidence of shifting settlement in Eastern Europe. It is only following the end of the Younger Dryas cold event that the first occupations appeared in newly deglaciated areas of the eastern Baltic—in Estonia and Latvia—between roughly 56° and 60° North. These sites are assigned to the *Kunda* culture and dated to 11,500–7,000 years ago. They contain tanged points and other stone artifacts similar to those of the early postglacial sites in northwest Europe.[37]

Sites were also occupied in the interior of Russia up to latitude 63° North during 10,000–9,000 years ago, and—combined with the Kunda culture sites—they provide a picture of expansion into the northern forest. In the wake of deglaciation, these regions were rapidly colonized by pine and birch forest. The terrain was covered with lakes and bogs. Climates were slightly cooler and drier than those of the present day, and the January mean was probably below 0° F (−17° C) in the interior of northern Russia. A boreal forest fauna, including moose *(Alces alces)*, aurochs, wild boar, and beaver, had replaced mammoth, horse, bison, and other mammals of the periglacial plain, although reindeer roamed the Far North.[38]

Buried in bog deposits that ensured exceptional preservation of bone and wood, sites like *Zveinieki* in Latvia and *Nizhnee Veret'e* and *Vis* in northern Russia provide a rare glimpse of the true complexity of material culture in cold environments at this time. They offer a sharp contrast to the sparse collections of stone artifacts from Scandinavian sites of the same period, described earlier. The north Russian sites in particular document what was probably the first human adaptation to subarctic boreal forest or northern taiga. The diversity and complexity of the technology—including many items of wood, bark, and other plant products—is reminiscent of the northern Athapaskans in interior Alaska, who occupied a similar setting.[39]

Most of the artifact types found in the Epi-Gravettian can be identified in these sites (for example, needles and needle cases), but many novel forms of technology are also present and a new burst of innovation seems to have taken place. A variety of mechanical devices were produced. The innovations were almost certainly part of the radical shift to a

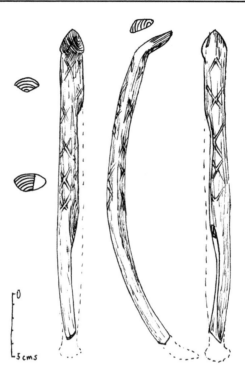

Figure 6.4. Small wooden bow from the site of Vis I in northern Russia.

northern forest economy, which seems to have been achieved within a few centuries.[40]

Wooden bows (pine and spruce) and arrows are well represented, and the latter include both blunted forms for hunting birds and long pointed forms for large mammals. Some of them were fletched. A series of short bows—apparently designed for operating drills and/or fire making—were recovered from Vis I. Pike and other fish were caught with harpoons and multipronged spears, or leisters. Landing nets or fish traps were built with hoops, and floats were made of pine bark.[41]

A revolution in transportation technology also may have taken place. Fragments of wooden skis and sledge runners were found at Vis I, along with at least one oar. Domesticated dogs are associated with these sites, but it is not clear if they were used to pull sledges over snow and ice.[42] The ability to move people and materials quickly and easily across the landscape, along with the new hunting and fishing technology—which seems to have included at least some untended facilities—must have increased the efficiency of the food-getting economy.

Despite these impressive technological achievements, analysis of the wooden long bows reveals a primitive design. In later millennia—during the Neolithic and Bronze Age—bow design was refined for better performance.[43] Progressive improvement of technology is a pattern well established in later prehistory and history, but more difficult to document in the remote past. It may have played an important role in the settlement of higher latitudes, where technology was so critical to survival. Gradual improvement of a number of technologies developed by modern humans (such as insulated fur clothing) could have been an important factor in the colonization of the coldest regions of northern Eurasia, although this cannot be confirmed because of the limited preservation of these technologies in the archaeological record.

It is not known if the occupants of these sites were spending lengthy periods at one location, but they were constructing houses with a variety of carved wooden components—pegs, stakes, blocks, and beams. Houses were rectilinear in plan and occupied an area of 120–150 ft^2 (40–50 m^2). Hints of a less nomadic lifestyle are provided by a profusion of household goods, such as birch bark containers and vessels, wooden cutting boards, fish-cleaning implements, fiber mats, and others.[44]

For the first time, people were burying their dead in cemeteries—not the isolated graves of the Upper Paleolithic—and this also points toward a more sedentary society. Especially well known is *Oleneostrovskii* on Lake Onega, which was initially investigated in 1936–1938, yielding more than 170 burials of men, women, and children. Grave goods and other evidence of ritual are common. At the smaller cemetery of *Popovo*, complete fish skeletons were found buried alongside adult males.[45] Other expressions of worldview can be observed in wood sculptures of people and animals. They are all very different in content and style from the Epi-Gravettian art and suggest that this worldview was changing along with the technology, organization, and environmental setting.

The analysis of skeletons from Oleneostrovskii and other postglacial cemeteries in northern Europe reveal new anatomical adaptations to cold climate in the form of shortened distal limb segments (that is, reduced brachial and crural indices). Similar changes are not apparent in southern Europe. This was a trend that continued among the later peoples of the circumpolar zone.[46]

Some sites are found north of the boreal forest in the tundra zone and might have been occupied—at least on a seasonal basis—during this interval. They are found in the northern Pechora Basin, slightly above the Arctic Circle (66° North) and near the coast of the Barents Sea.[47] Although the remains are confined to surface scatters of stone artifacts and their age is uncertain, the presence of well-dated sites further east along the Siberian Arctic coast suggests that people were probably here as well.

Siberia

Siberia also remained largely uninhabited during the cold maximum of the Last Glacial. It seems to have been reoccupied after 21,000 years ago—more or less simultaneously with the East European Plain. Several sites in southern Siberia date to between 21,000 and 19,000 years ago (for example, *Studenoe* in the southern Lake Baikal region), but most sites of the post–cold maximum epoch are younger. The latter include localities like *Kokorevo* in the Yenisei Valley and *Chernoozer'e* in the Ob' River Basin. They are confined to latitudes below 57° North and most are radiocarbon-dated to between 19,000 and 14,000 years ago.[48]

The Siberian sites are rather different from their Epi-Gravettian counterparts and they seem to reflect a more mobile way of life. Groups of mammoth-bone houses, storage pits, and other signs of long-term settlement are absent. Occupation areas are comparatively small. Traces of dwelling structures are rare and limited to small temporary shelters (such as circular tent rings). Examples of visual art are uncommon.[49]

Unlike the central East European Plain, southern Siberia supported some trees—chiefly pine—during this period. This is indicated not only by pollen-spore data but also by the presence of wood charcoal in former hearths at archaeological sites. Although reindeer dominate the mammal bones in these sites, some of them contain woodland animals such as red deer *(Cervus elaphus)* and occasionally even moose.[50] The use of wood for fuel and material could account for some of the contrast with the Epi-Gravettian sites.

The primary source of the difference, however, would seem to be that the Siberian habitat simply did not provide opportunities to concentrate food resources in specific places on a regular basis. The large mammal biomass was probably lower than that of the East European periglacial steppe, and periodic concentrations of fish or other smaller game were apparently not available. Accordingly, the population was forced to remain dispersed and mobile throughout the year—small groups constantly on the move.[51] A marriage network must have been maintained through periodic contacts among families.

The technology of the Siberians also seems to reflect this highly mobile existence. Much of their stone technology was focused on the mass production of tiny blades from characteristic wedge-shaped cores. The microblades were no more than ¼-inch (4 mm) wide with unusually sharp edges. They were fixed into grooves along one or both edges of a sharpened bone or antler point. Specimens of complete microblade-inset points have been recovered from both Kokorevo and Chernoozer'e. At Kokorevo, one was found embedded in a bison shoulder blade.[52]

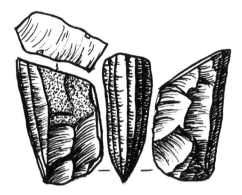

Figure 6.5. Wedge-shaped microblade core from Dyuktai Cave on the Aldan River in northeastern Siberia.

Microblade technology is an extremely efficient use of high-quality stone. It generates the maximum amount of usable edge while minimizing the quantity of stone that must be gathered and carried around. The small wedge-shaped cores often doubled up for use as scraping or cutting tools, further illustrating the efficiency and frugality of the Siberians.[53] In fact, microblade technology—and more generally "microlithic" tools—are broadly associated with cold climates, although they sometimes show up in other places, such as Mexico.

Aside from the grooved points, the diversity and complexity of bone, antler, and ivory artifacts seem poor in comparison to the Epi-Gravettian. Large and small needles are present, along with awls, thin points, hide-burnishers, and wedges and/or stakes.[54] Some implements might have been fashioned out of wood, which is not preserved in any of these sites. But much of the contrast probably is due to a less diverse economy—based chiefly on the hunting of large mammals—with limited procurement of birds, fish, and other animals that require complex food-getting technology. However, some small mammals (especially hare) were exploited for food and/or fur.[55]

Human skeletal remains are very scarce in these sites, and it is not clear if the Siberian population had evolved the same anatomical adaptations to cold that are observable among the Europeans after the cold maximum. Some upper-limb bones were recovered many years ago from *Afontova Gora* in the Yenisei Valley, but they were too fragmentary for measurement of the length of the arm or its segments.[56]

As climates warmed after 15,000 years ago, some changes in economy took place. Remains of fish and barbed harpoons are present in several south Siberian sites—most notably at *Verkholenskaya Gora* on the Upper Angara River—and forest-dwelling mammals become more common.[57] A major transition is evident at the nearby site of *Ust'-Belaya* by 11,000 years ago. The large area of this site suggests increased settlement size, and debris-filled pits are reported. The fauna, which is entirely modern (for example, roe deer and moose), includes remains of fish and traces of domesticated dog. Some new technology is indicated by the appearance of fish hooks—along with the barbed harpoons—among the bone and antler implements.[58]

People expanded northward into the Middle Lena Basin. By 15,000 years ago, they were visiting *Dyuktai Cave* on the Aldan River at 59° North. This small cave became the type site of the *Dyuktai* culture, defined by the colorful Russian archaeologist Yuri Mochanov more than thirty years ago. The artifacts are similar to those of the south Siberian sites and include wedge-shaped cores and microblades, along with some bifacial tools, burins, and scrapers. The mammal bones found with the artifacts include extinct periglacial forms (such as steppe bison), but forest taxa appear in the upper levels.[59] Dyuktai is of special interest to American archaeologists because it probably represents the first culture to spread across the Bering Land Bridge and into the New World.

A new culture materialized roughly 12,000 years ago and quickly dispersed over large portions of northern and eastern Siberia. Also defined by Mochanov, the *Sumnagin* culture is an enigma.[60] Although contemporaneous with the boreal forest settlement of subarctic Europe, Sumnagin sites are small and yield a comparatively limited array of artifact types. The stone tools are completely dominated by items made on small blades struck off thin cylindrical cores (often characterized as *karandashevid'nii*, or "pencil-like" cores, by Russian archaeologists). Implements of bone and antler are scarce—barbed harpoons and fish hooks are absent.[61]

Some of the contrast between Sumnagin and its European counterparts is due to the lack of wood preservation. Spectacular bog discoveries like Vis I are unknown in Siberia. Most Sumnagin sites were located in the forest zone and some of the technology (and perhaps all of the art) was undoubtedly created from wood. But—as during the period following the cold maximum—most of the differences with north European sites probably lie in the relatively low productivity of the Siberian environment. The boreal forest of the Middle Lena Basin today supports roughly half the biomass of the subarctic forest at the same latitude in northwest Russia.[62]

The consensus among Russian archaeologists is that the Sumnagin people subsisted primarily on a diet of large mammals. Moose predominates among the faunal remains, followed by roe deer, reindeer, and brown bear. With the possible exception of reindeer, all of these mammals are found at low densities in subarctic forest. The remains of fish and water birds are rare in Sumnagin sites, despite their invariable location along river margins.[63]

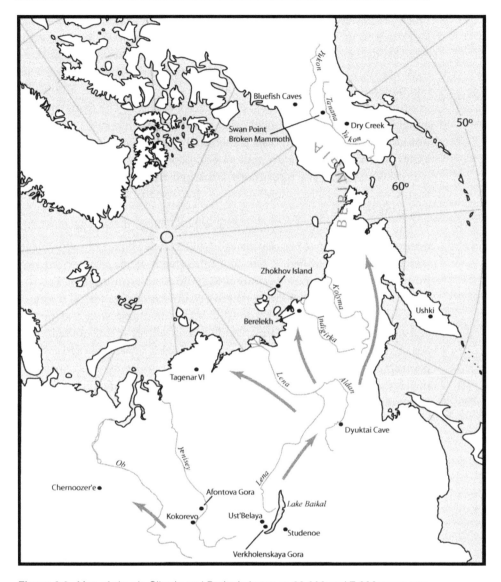

Figure 6.6. Map of sites in Siberia and Beringia between 20,000 and 7,000 years ago.

Nevertheless, the Sumnagin culture (or something very like it) spread northward into the tundra zone after 10,000 years ago and became the first to occupy the Siberian Arctic. Temperature, moisture, and biological productivity are all generally lower here than in the boreal forest. The amount of solar energy reaching the surface of the earth on the Arctic Circle—roughly 66° North—is about half of the amount received on the Equator. At the same time, the loss of solar energy due to reflected light *(albedo)* is much greater at high latitudes due primarily to the presence of snow and ice. Much of the ground is permanently frozen *(permafrost)*. Little wood is available except in the form of driftwood along some coastal margins.[64]

A site on Zhokhov Island at 76° North between the Laptev and East Siberian Seas is dated to 9,500–9,000 years ago. At the time, the island was probably joined to the coast or a larger land mass. The stone artifacts are quite similar to the Sumnagin sites of the forest zone, but a number of nonstone implements are also present. They include slotted bone and antler points, antler and ivory mattocks, and bone handles for cutting tools. A few wooden artifacts (apparently fashioned from driftwood) were also found, including a large shovel or scoop, arrow shafts, and a sledge-runner fragment.[65]

The animal remains recovered from Zhokhov Island are primarily reindeer and polar bear *(Ursus maritimus)*. Only isolated bones of walrus, seal, and birds were identified.[66] The first inhabitants of the arctic coast had brought an interior hunting economy with them. An economy based on the much richer resources of the northern ocean would take time to develop.

Sumnagin spread eastward into Chukotka, and also westward along the Siberian arctic coast to the Taimyr Peninsula on the other side of the Laptev Sea. An occupation at *Tagenar VI* is dated to roughly 7,000 years ago and apparently represents the earliest-known human presence here. According to the analysis of pollen samples, the forest zone had expanded into the area at the time and climates were slightly warmer than today.[67]

Beringia and the New World

In central Alaska, gold miners—not paleontologists—have recovered the largest quantities of Pleistocene animal remains. Millions of bones of mammoth, bison, horse, and other mammals that roamed the Alaskan landscape were buried in massive silts deposited during and before the Upper Pleniglacial. Blasting the silt with high-pressure hoses, miners washed out vast numbers of these bones along with the gold. Some of them were still attached to shreds of dried and frozen skin and hair, and occasionally a partial carcass has been found.[68]

Examining the great quantities of remains—and noting the preponderance of large grazing mammals such as steppe bison and horse—the paleontologist Dale Guthrie concluded in the 1960s that Alaska must have supported a rich grassland prior to the end of the Pleistocene. This view was at variance with the herbaceous tundra habitat reconstructed by pollen specialists and sparked a lively debate. Over the past few decades, radiocarbon dates on the bones have accumulated, and it is now clear that—despite its significantly higher latitude—central Alaska contained most of the same biota found in the periglacial steppe environments of northern Eurasia during the Upper Pleniglacial.[69]

The key to this seeming paradox lies in the increased aridity that prevailed over the land between the Lena Basin and the northwest Canada at the time. Sea level had fallen almost 400 ft (120 m) below its present position, exposing a dry plain between Chukotka and western Alaska (named *Beringia* by the Swedish botanist Eric Hultén). Clear skies, reduced precipitation, and loess deposition promoted well-drained, nutrient-rich soils that supported diverse steppic plant communities—no trees—and herds of large grazing mammals. The wet tundra soils and spruce bogs so common today were absent.[70]

The wealth of food resources raises the question of why humans did not colonize Beringia earlier. The oldest known sites are between 15,500 and 14,000 years ago, and date to a time when that wealth was actually declining as moisture and trees returned to the region. The problem for modern humans was their apparent inability to occupy areas of Siberia above latitude 60° North until after 16,000 years ago. This effectively blocked earlier movement into Beringia, regardless of its abundance.

The settlement of Beringia—and by extension of the New World—was a consequence of the initial reoccupation of Siberia 20,000 years ago, followed by northward expansion into the Middle Lena Basin as climates warmed after 16,000 years ago. If the Siberian population had evolved some of the same anatomical adaptations to cold climate as their European counterparts in the wake of the cold maximum, this might have played a role in their colonization of areas above 60° North. Improvements in technology—especially in clothing and shelter—also could have been an important variable. Beringian settlement coincides with the return of woody shrubs and trees, and the introduction of a wood fuel source after 16,000 years ago might have been another factor.[71]

In any case, people were camping along the Tanana River in central Alaska by 14,000 years ago, and may have visited *Bluefish Caves* in the Yukon as early as 15,500 years ago. The lowest occupation levels at the Tanana Valley sites (latitude 64° North) contain artifacts similar to those of the Siberian Dyuktai culture. At *Swan Point,* these comprise microblades, burins, and flakes struck from bifacial tools. Artifacts at the nearby site of *Broken Mammoth* are few, but include several rods of mammoth ivory. In contrast to Dyuktai, the Alaskan occupations indicate a broad-based northern diet of large mammals, birds—chiefly waterfowl—and at least some fish. The avian fauna is diverse and includes ptarmigan, mallard, pintail, teal, swan, and various geese.[72]

Figure 6.7. Excavation of the Broken Mammoth site on the Tanana River in Alaska.

Access to other parts of the New World may have been denied during this period. Massive ice sheets covered most of Canada and the Northwest coast. An "ice-free corridor" through western Canada to the northern Plains is thought to have opened up no earlier than 13,500 years ago. However, deglaciation in the Pacific Northwest now seems to have taken place more rapidly, and a coastal route could have been available by 17,000 years ago. Sites are firmly documented in midlatitude North America by 13,500 years ago, and at least one site in South America may have been occupied as early as 15,000 years ago.[73]

Rising temperatures and increased moisture accelerated environmental change after 14,000 years ago, as shrub tundra replaced steppic habitat in many parts of Beringia. Most of the large grazing mammals of the Last Glacial Maximum had become extinct.[74] Some changes are evident in the archaeological record, although their meaning is not entirely clear. At the lowest levels in the *Ushki* sites of central Kamchatka (recently redated to about 13,000 years ago), people built small oval houses and manufactured a stone tool kit that included stemmed bifacial points. Stone pendants, beads, and a burial pit are present, but little was preserved in the way of nonstone artifacts or animal remains.[75] The relation of these occupations to the Dyuktai culture and other Beringian sites is unknown.

In central Alaska, people moved up into the northern foothills of the Alaska Range to hunt elk and sheep at least on a seasonal basis (probably during the autumn). The earliest levels at *Dry Creek* and other sites in the Nenana Valley are dated to 13,500–13,000 years ago. These sites contain small bifacial points, but no bone or antler artifacts were preserved.[76] As in the case of Ushki, their relation to other Beringian sites is unclear. Because the ice-free corridor was open at this time, some archaeologists have speculated on links with early sites on the northern Plains, but this is controversial.[77]

Beringia remained intact through most of the Younger Dryas cold event, which began after 13,000 years ago. Sites containing wedge-shaped microblade cores and microblades, along with burins and bifacial tools, are found in central Kamchatka (Ushki), many parts of Alaska, and probably Chukotka. Their stone artifacts, like those of most of the earlier Beringian sites, are similar to those of the Dyuktai culture. Unfortunately, only isolated nonstone implements are preserved.[78]

The Younger Dryas period sites reveal a broad-based economy and diet like that of the early Beringian sites. Fish bones are common in the former hearths at Ushki, and the occupation level dating to this interval at Broken Mammoth in central Alaska contains the same inventory of birds, fish, and mammals found in the lowest layer. These sites also yield evidence of technological responses to cold climate, which may reflect the sudden return of lower temperatures at this time. Traces of former houses at Ushki exhibit shallow entrance tunnels—apparently the oldest-known examples on earth—designed to keep cold air out of the dwelling space. Broken Mammoth produced an eyed needle of bone.[79]

At some point roughly 12,000 years ago, the rising sea level reached a position less than 150 ft (50 m) below today and flooded the lowlands between Chukotka and western Alaska. This increased available moisture over the remaining land areas and accelerated the transition to wet tundra and coniferous forest. Beringia had ceased to exist.[80]

During the next few thousand years, the archaeological record of Alaska probably reveals less change than any other part of the world. People continued to produce small wedge-shaped microblade cores, burins, and bifacial implements of stone. Isolated examples of bone artifacts—such as a slotted point—have been recovered at a few sites, but generally little is known about their nonstone technology. Many of the Alaskan sites are considered part of the local *Denali* complex or culture (roughly 12,500–7,500 years ago).[81]

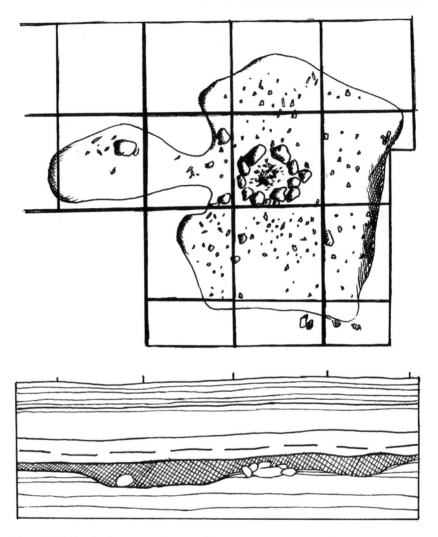

Figure 6.8. Semi-subterranean house with entrance tunnel at Ushki on Kamchatka.

Denali sites are small and—like the Siberian Sumnagin sites—suggest a dispersed population and mobile way of life. Poor preservation of animal remains, however, has obscured much about the diet and economy. Caribou, sheep, and moose bones are found in some sites, while information on fish and bird procurement is scarce. Site numbers decline during the warm phase that followed the Younger Dryas cold event, but increase during a subsequent cool oscillation about 8,000 years ago. The pattern might reflect a response to improved conditions for caribou hunting.[82]

As in Siberia, there are no signs of the arctic maritime adaptation that would one day dominate the Far North. Although some Denali sites have been found in northern Alaska, they are unknown along or near the northwest and northern arctic coast.[83] In contrast, large coastal occupations and exploitation of marine resources are well documented on the North Pacific rim (specifically on the Aleutian Island chain) and along the coast of southeast Alaska.[84]

CHAPTER 7

Peoples of the Circumpolar Zone

One of the most profound consequences of the climate warming that ended the Ice Age was the creation of latitudinal vegetation zones across northern Eurasia. During the last glacial period (75,000–12,000 years ago), environments were more homogeneous, as illustrated by the fact that most of the same large mammals may be found in deposits of this age between 40°–60° North (and even further north in Beringia). Although some latitudinal—as well as longitudinal—variation is evident, it was subdued in comparison to the sharply delineated zones that developed after 12,000 years ago.[1]

The northernmost zone is the tundra, which coincides broadly with the Arctic Circle at 66° North, although coniferous forest extends above this latitude in some regions. The tundra is cold and dry, lacks trees, and supports a characteristic array of plants and animals—many of which are confined to high latitudes. While tundra is one of the least productive of terrestrial habitats, subarctic and arctic waters contain some highly productive environments—especially where "upwelling" zones exist. In such areas, rising currents bring nutrients up from the ocean depths to the surface, supporting large populations of fish and sea mammals.[2]

If the widespread steppe and forest-steppe habitat of the glacial period allowed for an unusual degree of cultural homogeneity over large regions, the marked environmental zonation of the postglacial period promoted increased cultural diversity. Sharp differences in language and culture emerged among the peoples of the Eurasian steppes, deciduous woodland, boreal forest, and other zones. The peoples of the Far North became especially isolated from other cultures because of their remote geographic position and harsh climate setting. It was impossible to establish a farming economy in most boreal forest and tundra environments (although reindeer herding developed eventually in many parts of northern Eurasia).

The cultural isolation of what had now become the "cold frontier" of human settlement is one of the more striking patterns of circumpolar prehistory. The pattern is most apparent in the North American Arctic, but even in Western Europe—close to evolving centers of empire and industry—people living at high latitudes developed and maintained a very separate way of life. The extent to which changes in the archaeological record of the Arctic can be attributed to influences from the south has been a subject of some debate among researchers—particularly in Europe.[3]

The most important trend in circumpolar prehistory over the past 7,000 years was the development of an arctic maritime economy. Only by overcoming the formidable technical challenges posed by hunting sea mammals in open water and under polar ice could people exploit the great wealth of marine resources in the Far North. And in this case—regardless of other influences from outside cultures—the key technological innovations were achieved independently by northern peoples. Arctic maritime economies did not develop everywhere, however, and there was much variety in how different peoples responded to these challenges. The prehistory of the circumpolar zone illustrates as well as anywhere the unique historical destiny of each language and culture.[4]

The final chapter in the human settlement of higher latitudes began 7,000–6,000 years ago with a major episode of global warming (often referred to as the *Atlantic* period). Temperatures in the northern hemisphere eventually reached levels not seen for more than a hundred thousand years. The effect on human populations in the circumpolar zone was significant, and major changes are apparent in the archaeological record. The largest impact fell on North America, where accelerated deglaciation opened up the Canadian Arctic and Greenland for rapid settlement.

The European Arctic

As we have seen, modern humans expanded above the Arctic Circle in many deglaciated areas of Eurasia (including Beringia) between 12,000 and 7,000 years ago. However, even in areas where they occupied coastal settings (such as Zhokhov Island), there is no evidence for significant use of marine resources. Only in milder subarctic settings with especially high marine productivity—such as the Aleutians and southeast coast of Alaska—did maritime adaptations emerge before 7,000 years ago.[5]

The exception to this pattern may be the northwest European Arctic, where the Komsa culture had established itself at some earlier point (see chapter 6). Although the archaeological record provides little information on the Komsa economy, some have argued that it probably represented an early adaptation to marine resources.[6] It may be important that—owing to the effects of the Gulf Stream—the arctic coast of Norway is rather similar to the subarctic coastal settings mentioned above in terms of climate and marine resources. If the Komsa people had developed a maritime economy, the changes observed in the archaeological record after the onset of the Atlantic period may have had less significance than is sometimes thought.

European archaeologists have designated the period between 7,000 and 2,000 years ago as the *Late Stone Age* in arctic Norway. Although several

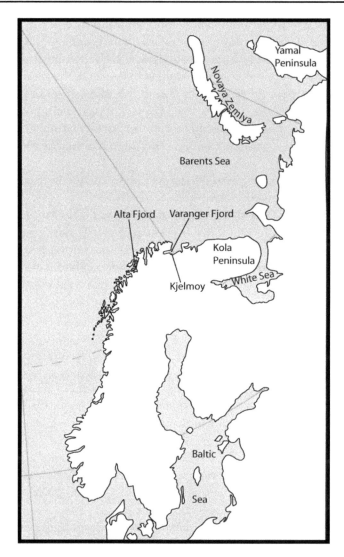

Figure 7.1. Map of sites in the European Arctic.

phases are distinguished on the basis of changes in artifacts and features, there is overall continuity within this time period. Throughout the Late Stone Age, pottery and a variety of ground slate implements were present. Chipped-stone tools and weapons were also present, although the microblade technology that appeared in late Komsa times was not. And rock art in the form of engravings—which reveal much about Late Stone Age life in the European Arctic—became common.[7]

Whatever its antecedents, the Late Stone Age economy was firmly based on the heavy use of marine resources—supplemented with some terrestrial foods. The analysis of bone debris from sites in Varanger Fjord indicates that fishing was concentrated on cod, salmon, and halibut. Sea mammals—including seal, walrus, porpoise, and small whales—were hunted along with polar bears, while reindeer were hunted in the interior. Birds also contributed to the diet. Like other maritime economies, this one probably offered a relatively stable food supply.[8]

Sites are found in the fjords and on islands, and they contain two types of semi-subterranean houses. The smaller pithouse (*Karlebotn* type) occupied 20–50 ft^2 (7–15 m^2) with a single stone-lined hearth. A larger type *(Gressbakken)* was present between 5,000 and 3,500 years ago. It covered an area of 100–160 ft^2 (30–50 m^2) with two large interior hearths, and is typically associated with a midden, or refuse heap, composed of shell and bone. Some sites contain large numbers of houses, although it is apparent that many were constructed at different times. Nevertheless, the pattern suggests extended occupations—at least a limited form of sedentism—and some archaeologists have argued for a high degree of social complexity.[9]

The hunting and fishing technology included harpoons, gaffs, leisters, and fishhooks along with the chipped and ground stone points, knives, and other implements. The grinding and polishing of slate is an especially characteristic stone technology among arctic maritime peoples, and it eventually spread across the circumpolar zone. Many household items—such as bone needles and combs—are also typical of arctic cultures. Late Stone Age people developed (or perhaps inherited from their Komsa predecessors) transportation technology in the form of skis and boats—for which much of the evidence is derived from the rock art.[10]

Indeed, these people became their own ethnographers and carefully recorded their hunting and fishing activities, material culture, and ritual practices in rock engravings. Thousands of engravings are found in Alta Fjord. In addition to boats and skis, they depict moose, reindeer, and various birds. Geometric designs were also created.[11]

People living in the European Arctic were subject to more contact and influence from southern cultures than other peoples of the circumpolar

Figure 7.2. Rock carving depicting figure on skis from the Late Stone Age of northern Norway.

Figure 7.3. Saami fish weir.

zone. This was increasingly due to their proximity to expanding centers of power in southern Norway, Sweden, and elsewhere, but the comparatively mild climate might have been another factor. During the Late Stone Age, pottery was apparently introduced from foreign sources. The earliest form (*Comb* ceramic style) appeared more than 6,000 years ago and seems to be derived from southern Finland. After 4,500 years ago—following an interval during which ceramics became scarce—pottery tempered with asbestos appeared, and it too is believed to have southern origins.[12]

Pottery from Finland also spread into the northernmost parts of European Russia about 6,000 years ago. A maritime economy seems to have been established here somewhat later than on the arctic coast of Norway. By 4,500 years ago, peopling were hunting seal and whale along the shores of the White Sea. The diet was supplemented to some degree by moose hunting and freshwater fishing in interior forest areas. The technology apparently included large boats, which are depicted in local rock art.[13]

Major changes took place roughly 2,000 years ago, which marks the beginning of the local *Iron Age* in arctic Europe. At places like *Kjelmøy* (Varanger Fjord area), the first iron artifacts materialized alongside the familiar Late Stone Age technology. Bronze artifacts and evidence of local bronze casting may show up even earlier in northern Russia. Reindeer herding and some plant cultivation were also practiced after 2,000 years ago, although there is some evidence that they also date from earlier times.[14]

From the beginning of the Iron Age onward, the ethnic identity of the people living on the arctic coast of Norway and northwest Russia is clear. They are the Saami people (formerly referred to as the Lapps), who still inhabit the region today and are probably descended from the Late Stone Age population. Further east—along the coast of the White Sea and the Barents Sea—live the Nenets people, who also continued to pursue a maritime economy (including some whaling) during the historic period.[15]

As elsewhere in the circumpolar zone, the isolation and late survival of indigenous cultures provides a window on the prehistoric past. Historical accounts, ethnographic studies, and the traditional knowledge of the Saami and Nenets not only fill the gaps in the archaeological record pertaining to economy and society, but also inform us about the way they modeled and interpreted the world with words and other symbols.

Siberian Neolithic

Russian archaeologists in Siberia employ the same classificatory framework as their counterparts in Europe. The remains of past peoples are classified in accordance with a sequence of stages such as Late Stone Age (or Neolithic), Bronze Age, and so forth. This framework was developed in Western Europe during the nineteenth century and reflects the then prevailing view that all cultures progress through a sequence of evolutionary stages.[16]

The archaeological record of the Arctic—and perhaps the Subarctic as well—is not especially amenable to this framework. Even where progressive change may be evident, it does not follow the stages perceived in midlatitude Europe, and its archaeological correlates do not have the same meaning. In the traditional framework, the appearance of pottery marks the start of the Neolithic and is linked to settled village life and agriculture. But in the circumpolar zone, ceramic technology is tied to other developments. In fact, arctic peoples seem to have been somewhat ambivalent about pottery and some of them stopped making it at times.[17]

Nevertheless, pottery appeared in the Middle Lena Basin roughly 7,000 years ago—during the warm Atlantic period—at the beginning of what is termed the Siberian Neolithic. The surfaces of the clay pots were marked with net impressions and the rims were often notched to resemble an encircling cord. This pottery is associated with the *Syalakh* culture, which is found at sites like *Bel'kachi I* on the Aldan River. The Syalakh people

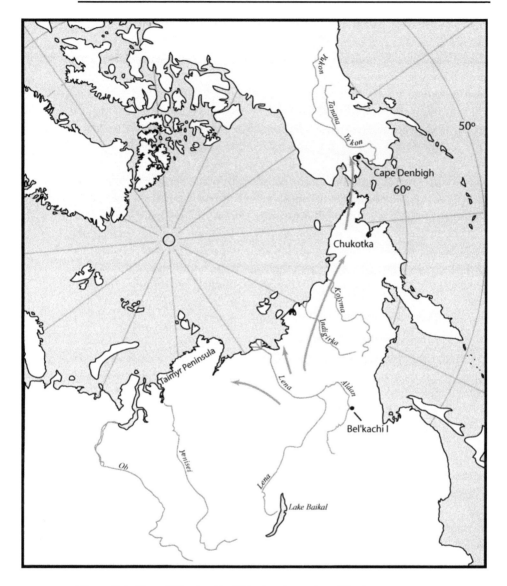

Figure 7.4. Map of Siberian Neolithic sites in northern Siberia.

produced a set of stone artifacts similar to those of their Sumnagin predecessors—cylindrical cores and small blades—but with the addition of delicately flaked bifacial points and ground adzes. Nonstone implements included barbed points of bone.[18]

The Syalakh culture spread quickly to the Siberian arctic coastal region and soon expanded to fill the same geographic area occupied by the earlier Sumnagin culture. Characteristic net-impressed pottery fragments can be found from the Taimyr Peninsula in the west to Chukotka in the east.[19] The rapid spread of people and/or culture traits around portions of the circumpolar zone was a pattern that would be repeated in later millennia.

Despite the Syalakh presence in the arctic coastal region and Chukotka, their sites do not indicate any shift toward the use of marine resources. The economy remained focused on subarctic forest habitat, and it should be kept in mind that the forest had expanded northward into areas of former tundra at this time. The sites are small and filled with the remains of moose and reindeer—the latter especially in tundra settings. However, an increased reliance on fishing in lakes and rivers is evident[20] and is perhaps related to the introduction of pottery.

About 6,000 years ago Syalakh was superseded by the *Bel'kachinsk* culture. A new form of pottery appeared—cord-marked with a cord-wrapped paddle. Production of small stone blades and finely flaked bifacial points continued, along with ground adzes. The sites are found along rivers at stream confluences, and the economy maintained a focus on hunting moose and other land mammals, along with fishing and some exploitation of birds.[21]

Like its predecessor, the Bel'kachinsk culture spread north into the Siberian Arctic and occupied the same area between Taimyr and eastern Chukotka. It seems to have expanded further, however, into northeast Chukotka and along the coast of the Sea of Okhotsk.[22] And Bel'kachinsk people—despite their interior focus—apparently crossed the waters of the Bering Strait and established themselves in western Alaska.[23]

In 1948 the American archaeologist J. Louis Giddings uncovered finely flaked bifacial points, microblades, ground adzes, and other typical elements of the Siberian Neolithic stone assemblages at a site on Norton Sound in Alaska. This site and others like it were assigned to the *Denbigh Flint complex* and dated to as early as 4,500 years ago—toward the end of the Bel'kachinsk period. Although pottery was not found in these sites, Giddings and others

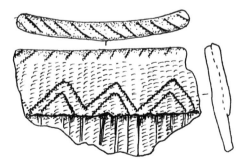

Figure 7.5. Fragment of cord-marked pottery (Bel'kachinsk culture) from the site of Sumnagin II.

recognized the link to the Siberian Neolithic, which more recent discoveries have strengthened.[24] Denbigh became the wellspring of cultures that eventually stretched across the North American Arctic, and its appearance in Alaska was a major event.

Pottery style changed again in the final phase of the Siberian Neolithic (*Ymyyakhtakh* culture), which began roughly 4,500 years ago. The new pottery was check-stamped. It was accompanied by some minor changes in stone tool technology, and imported items of bronze—then being manufactured in southern Siberia—are sometimes found in the sites. On the whole, however, no significant shifts in technology, economy, or society took place. When the Siberian Iron Age began about 2,500 years ago, the peoples of northernmost Europe represented the only arctic maritime adaptation in Eurasia.[25]

The Paleo-Eskimo World

At the beginning of the Atlantic warm period 7,000 years ago, most of the North American Arctic remained uninhabited. Large glaciers still covered portions of the Canadian Arctic, although they were shrinking. East of the Bering Strait, human settlement of the Arctic was confined to Alaska and northwest Canada, although Indian groups had moved into subarctic areas of central and eastern Canada during the early postglacial epoch.[26]

At the beginning of the Atlantic period, an entirely new culture appeared in Alaska. Artifacts of the *Northern Archaic* tradition are found throughout the interior of the state and sometimes on the coast. Especially characteristic of this tradition are side-notched stone points, although other types of bifacial points become common later. Most Northern Archaic sites are found in interior forest settings and preservation of nonstone objects is rare. Accordingly, little is known about the technology—which was probably based primarily on wood and bone—or the economy. The latter was almost certainly oriented toward boreal forest resources.[27]

In many respects, both the Indian groups that occupied central Canada and the Northern Archaic tradition of Alaska were similar to the Siberian Neolithic cultures described earlier. All of them seem to represent peoples who had adapted to an interior northern forest habitat after 12,000 years ago, and who subsequently moved northward as climates continued to warm during the early postglacial and Atlantic periods.

The Denbigh Flint complex arrived during the later phases of the Atlantic period, marking yet another of the many abrupt transitions in the Alaskan record. Most archaeologists believe that Denbigh represents the first shift toward an arctic maritime economy outside Europe. However,

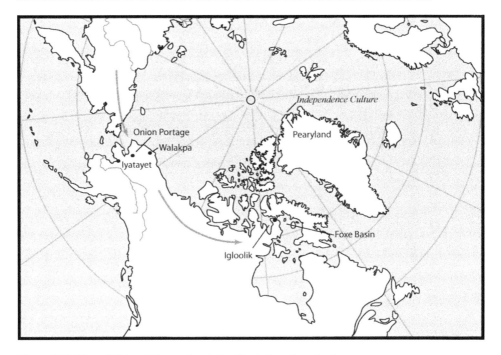

Figure 7.6. Map of Paleo-Eskimo sites in the North American Arctic.

though some burned seal bones were found at the original Denbigh site on Norton Sound *(Iyatayet)*, bone preservation is generally rare and most inferences about the diet and economy are based on the stone artifacts and information from related sites in the Canadian Arctic.[28]

In addition to a typical Bel'kachinsk assemblage of stone artifacts, Denbigh sites contain small end-blades and side-blades, both of which are later associated with harpoons for hunting sea mammals. At similar sites in the Canadian Arctic (described below) where bone objects were preserved, harpoon heads, lance heads, fish spears, and other marine hunting gear were found. The Canadian sites also yielded remains of seal and walrus.[29]

Denbigh sites are located on the west and northwest coast of Alaska, and a late occupation may be present along the northern coast at *Walakpa* (near Barrow). The dwellings in these sites are small and seem to be temporary—reflecting seasonal occupation. Other sites are found in the northwest interior at places like *Onion Portage* on the Kobuk River, and these contain traces of houses that were probably occupied during colder months. The interior sites were undoubtedly used for caribou hunting, and they underscore the mixed character of the Denbigh economy. Marine foods were exploited on a seasonal basis by groups that remained small, mobile, and highly dependent on land resources.[30]

Box 5. Climate Change in the Northern Hemisphere: The Last 7,000 Years

Since the end of the Pleistocene about 12,000 years ago, the northern hemisphere has experienced interglacial conditions with temperature oscillations that were relatively small in comparison to the extreme variations of the Middle and Late Pleistocene. Nevertheless, these oscillations had significant impacts on food production and population growth among agricultural societies of the temperate zone. And even small changes in mean annual temperature (2°–3° F or 1°–2° C) had major effects on settlement and economy in the subarctic and arctic zones.[a]

Sources of information on climate change during the past 7,000 years are abundant. Although an oxygen-isotope record is available from ice cores, dated pollen cores extracted from lakes and peat bogs provide the primary framework for this time range. Additional sources of data include plant macrofossils, beetle assemblages, molluscs, tree-ring data, and others. During the last few thousand years, written historical sources provide especially valuable information on past climate.[b]

The Atlantic period, which is dated to between roughly 7,800 and 5,700 years ago, marked the climatic optimum of the postglacial (Holocene epoch). July temperature means in the northern hemisphere rose as much as 3.5° F (2° C) above those of the present day. Warm temperatures continued to prevail during the succeeding two millennia, but after roughly 2,500 years ago, they began to fall as much as 2° F (1° C) below current levels. Climates during the last 2,500 years are sometimes termed the Sub-Atlantic period. However, a slightly warmer interval took place between roughly AD 1000 and AD 1450 (Medieval Warm Period). This was followed by the "Little Ice Age" (roughly AD 1450–1850), when mean annual temperatures dropped one or two degrees below current levels.[c]

---

a. See, for example, Hubert H. Lamb, *Climate, History, and the Modern World*, 2nd ed. (London, Routledge, 1995); Brian M. Fagan, *The Little Ice Age: How Climate Made History, 1300–1850* (New York: Basic Books, 2000).

b. See, for example, Astrid E. J. Ogilvie, "Documentary Evidence for Changes in the Climate of Iceland, A.D. 1500 to 1800," in *Climate since A.D. 1500*, ed. R. S. Bradley and P. D. Jones, pp. 92–117 (London: Routledge, 1992).

A year before Giddings found the first Denbigh site, the Danish archaeologist Count Eigil Knuth discovered a very similar set of remains in Pearyland, northern Greenland (at latitude 83° North). They were subsequently placed into the same cultural tradition *(Arctic Small Tool)* and recognized as the earliest evidence for the settlement of the central and eastern arctic regions of North America. Knuth named the new culture *Independence* (after

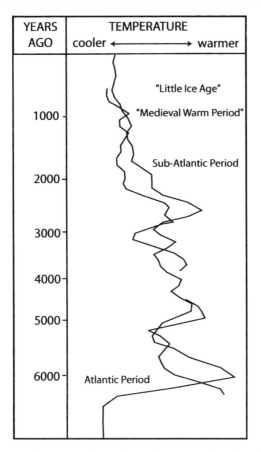

Figure B5. Temperature curve based on pollen cores from localities on Baffin Island in the Canadian Arctic.

---

c. H. E. Wright, Jr. et al., eds., *Global Climates since the Last Glacial Maximum* (Minneapolis: University of Minnesota Press, 1993); Neil Roberts, *The Holocene: An Environmental History,* 2nd ed. (Oxford: Blackwell Publishers, 1998).

Independence Fjord), and went on to reconstruct one of the strangest episodes in human prehistory.[31]

The Independence people had moved eastward to northern Greenland almost 4,500 years ago—about the same time that Denbigh appeared in Alaska (a slightly older "Proto-Denbigh" occupation is reported at Onion Portage). Although climates were warmer than present, Pearyland was a polar desert of rock, gravel, and ice. Complete darkness descended over the landscape during the long winter. Nevertheless, a variety of land and sea mammals inhabited the region, along with fish and migratory birds.[32]

Animal remains recovered from Independence sites reveal an economy that was fundamentally similar to that of the Denbigh people, but that reflected a High Arctic setting. Musk-oxen were hunted intensively. Smaller game, including polar fox, hare, and various birds, were also heavily exploited. Trout and char were fished, and at least some seal and walrus were hunted.[33]

The technology was rather limited in comparison to later cultures of the High Arctic, and Knuth concluded that the lives of the Independence people had been exceptionally harsh. Because caribou were not hunted, their clothing was presumably made from other, less suitable hides. Incredibly, they lacked the knowledge or ability to use sea-mammal oil for light and heat—lamps are absent. The hearths contain traces of scavenged driftwood, willow stems, and bone food debris, which probably restricted the number and duration of fires. In fact, Knuth speculated that they might have spent much of the winter period in a "kind of torpor" or state of hibernation.[34]

Roughly 4,000 years ago, the Independence culture disappeared from northern Greenland (although it later reappeared in modified form). At the same time, another Arctic Small Tool culture emerged in the central Canadian Arctic. Sites of the *Pre-Dorset* culture are found in the northern Hudson Bay and Foxe Basin region, where land and marine animal resources are especially rich.[35]

The Pre-Dorset people developed the first true arctic maritime economy of the western hemisphere, although many sites were located below the Arctic Circle—as far south as latitude 57° North. Seal and walrus apparently provided this people's primary food base, which was supplemented with caribou, fish, and birds. Sea-mammal hunting was probably critical to sustaining Pre-Dorset settlement through the normal fluctuations in terrestrial populations (which perhaps had ultimately doomed the Independence people).[36]

Although Pre-Dorset stone artifacts were similar to those of Denbigh (microblades, burins), nonstone technology was probably much more complex. The Pre-Dorset people carved toggling harpoon heads with sharp spurs and line holes from bone. These were designed to catch in the wound—detaching from the harpoon foreshaft—and allow retrieval of sea mammals in the water with the line. Some terrestrial hunting was done with a recurved composite bow and arrow. Fishing equipment included bone gorges and barbed points (possibly parts of leisters) and perhaps some of the fish weirs found near their sites. The remains of boats are unknown, but

Figure 7.7. Pre-Dorset harpoon head from the Canadian Arctic.

the distribution of sites across bays and islands suggests they may have been present. There is no evidence, however, of sleds and dog traction.[37]

In contrast to Independence, Pre-Dorset people employed sea-mammal oil for fuel, which they burned in small round lamps of soapstone. Lamps would have permitted habitation of domed snowhouses. Although no remains of flat snow knives have been found (used by later peoples to make these structures), circular patterns of debris on Devon Island and elsewhere have been interpreted as traces of snowhouses. The latter would have provided a means of camping on the winter ice for the hunting of ringed seal—a potentially vital resource. Constructed houses are not found in Pre-Dorset, and other dwellings were confined to tents with interior hearths (indicated by boulder rings).[38]

Climates had begun to cool after 4,000 years ago, and most Pre-Dorset occupation took place after the warm Atlantic period. Continued cooling after 3,500 years ago seems to have generated some culture change, and sites younger than 3,000 years ago are considered part of the *Dorset* culture. Large semi-subterranean winter houses were constructed and heated efficiently with lamps (not interior hearths). Snow knives, sled shoes, and other equipment associated with winter-ice hunting are found in Dorset sites. Curiously, bow and arrow use became rare, perhaps reflecting an intensified focus on sea-mammal hunting. There was a general expansion of technology in terms of diversity and sophistication.[39]

At the same time, art—which is almost completely absent in Pre-Dorset and Independence sites—became common. In fact, Dorset art is so rich and complex that scholars have attempted to deduce aspects of the worldview that it reflects. Sculptures of animals—bears, walrus, seals, birds, and others—were carved in both realistic and abstract styles. People and humanlike figures were also common subjects, along with rather eerie-looking masks. Much of the art was probably tied to magic beliefs and shamanistic practices. For example, human and bear carvings sometimes possess holes in the chest or throat into which slivers of wood were inserted.[40]

Climates began to warm slightly about 2,000 years ago, and the Dorset people reached a high point in terms of the numbers and sizes of their sites. Settlement expanded northward into the High Arctic, including northwest Greenland.[41] In some late Dorset sites (for example, *Igloolik*), large feature complexes comprising a central line of hearths enclosed by an arrangement of boulders up to 145 ft (45 m) in length are found. These feature complexes—often termed "longhouses"—are remarkably similar to those of the Gravettian on the central East European Plain (described in chapter 5). They were probably used in a similar way—to host temporary gatherings of families who were otherwise dispersed over large areas. These gatherings may have coincided with temporary concentrations of resources (such as walrus), but they almost certainly were used to reinforce social and economic ties among people. Long-distance movement of materials also underscores the development of networks over vast territories.[42]

Dorset and other cultures of the Arctic Small Tool tradition are widely classified as *Paleo-Eskimo* in order to draw a fundamental distinction between them and the Inuit (or Neo-Eskimo) peoples that followed.[43] Although the Paleo-Eskimo concept seems at first glance similar to the evolutionary stage concepts employed by archaeologists in the Old World such as Late Stone Age or Upper Paleolithic, this is actually not the case. Paleo-Eskimo is a higher-order classification of archaeological remains that have been grouped into cultures and phases without reference to progressive stages of development, reflecting the "culture-historic" approach of most New World archaeologists. It is an approach that avoids forcing the myriad cultures of the Arctic into evolutionary stages that never took place.

Roughly 1,000 years ago, the entire Paleo-Eskimo world came to an abrupt end. Although isolated elements of Dorset culture (for example, domed snowhouses) may have survived in some form—perhaps adopted by the later peoples of the Arctic—most of Paleo-Eskimo material culture vanished. The people who made it probably also vanished. They were replaced with a revolutionary culture that had emerged in the Bering Sea region during the preceding millennium and subsequently spread across much of the circumpolar zone.

Whalers and Warriors: The Thule Revolution

From the late 1500s onward, European explorers began probing into northern waters, initially around Greenland and eastern Canada but later between Siberia and Alaska. They eventually realized that the native people of Greenland spoke the same language as the people of the Bering Strait region. The longitudinal spread of Inuit—about 180° or halfway around the globe—is unequaled by any other language.[44]

The modern Inuit are the direct descendants of what archaeologists have termed the *Thule* culture (named after a settlement in northwest Greenland). Thule culture was the product of a number of interrelated technological and organizational developments that began in the Bering Sea region slightly more than 2,000 years ago. These developments enabled the

Alaskan ancestors of the Inuit to expand rapidly across the central and eastern Arctic after AD 1000, creating the remarkable uniformity of culture encountered by the Europeans.

Although the time and place of Thule origins are well known, the historical process that lay behind Thule's emergence remains unclear, and in fact constitutes what is perhaps the central problem of arctic prehistory. Thule culture had its beginnings in the so-called *Old Bering Sea* and *Okvik* cultures (or "phases") of Chukotka and St. Lawrence Island, which lasted from just prior to the start of the Christian era to about AD 700. But the underlying sources—in terms of people and ideas—have never been adequately explained. Some archaeologists believe that people living in western Alaska were the primary catalyst for the changes that eventually led to Thule. Others have argued that Northeast Asian influences were more significant. At this point, most would agree that the sources were varied and the process was complex.[45]

To begin with, there are hints of an older arctic maritime economy north of the Bering Strait on the Chukchi Sea. About 3,500 years ago, some people camped at *Cape Krusenstern* on Kotzebue Sound (Alaska). They built two sets of large houses—five for winter and five for summer—and left behind a mass of whale bones and stone blades that resemble later whaling harpoon heads. Giddings invented the term *Old Whaling* culture for these people, and both their origin and their fate remain a complete mystery.[46] Some years later, however, Russian archaeologists found a rather similar occupation on Wrangel Island (off the coast of northern Chukotka) dating to the same time period.[47] The relation of these sites to later developments is unknown.

After 3,500 years ago, changes took place in western Alaska that apparently reflect increased hunting of sea mammals. The Denbigh Flint complex was succeeded by a period during which people began using pottery and—at least in some sites—building large houses (*Choris* phase). The stone artifacts are similar to those of Denbigh, but also include ground slate knives and lamps. A variety of nonstone implements are preserved—barbed dart points, tanged arrowheads, spear rests, swivels, and others. Caribou were hunted at interior sites (such as Onion Portage), while seal and walrus were hunted on the coast.[48]

Choris was followed by the *Norton* phase, which began roughly 2,500 years ago. Defined by Giddings, who first encountered it at Iyatayet, the Norton phase is associated with a check-stamped pottery similar to that of the late Siberian Neolithic. The sites are relatively large—especially on Norton Sound—containing semi-subterranean winter houses with long entrance tunnels. Sea mammals were hunted with a simple toggle-head harpoon. A new emphasis on fishing is indicated by an abundance of notched-stone net sinkers, which may explain the sparse settlement of areas north of the Bering Strait (where lower water temperatures limit fish density).[49]

It is during the Norton phase that the Old Bering Sea/Okvik cultures emerge on the opposite shores of the Bering Strait. The sites are found along the coast of Chukotka (for example, *Cape Baranov*) and St. Lawrence Island, and their appearance marks a major expansion in technological complexity that includes critical innovations related to sea-mammal hunting. The toggling harpoon was perfected with a drag float apparatus, and the *umiak*—a large boat suitable for pursuing whales on the open sea—was developed. The technology included sealing darts used with throwing boards, bows and arrows, multipronged bird spears, hand-drawn sleds, and a profusion of other gadgets—the functions of which are sometimes unknown. The number of mechanical devices was high. Most chipped-stone items were replaced with ground slate, and the pottery style became simpler and cruder in appearance.[50]

The hunting of whales during the Old Bering Sea/Okvik period was limited, although settlements appear to have been located along whales' spring migration routes. The primary hunting focus was seal and walrus, with some attention to caribou and other terrestrial animals. The sites were of modest size, but contained semi-subterranean winter houses. Large cemeteries were established in at least two places on Chukotka—*Uelen* and *Ekven*—carrying significant implications for increased social complexity. The dead were buried with whalebones and numerous grave objects.[51]

In contrast to the Norton culture, the Old Bering Sea/Okvik people created an elaborate body of art. They carved figures of people and animals and also decorated harpoon heads, snow goggles, needle cases, and other equipment with complex geometric designs. The latter included curious "winged objects" that might have functioned as counterweights for harpoons. As in the case of the Dorset, much of this art was probably tied to magic practices and manipulation of the supernatural world.[52]

After AD 500, changes occurred in both the material culture and the distribution of settlement. Much larger sites—containing deep middens and accumulations of whalebone—appeared on coastal promontories of Chukotka and St. Lawrence Island (*Punuk* phase). These are associated with a growing emphasis on whale hunting. The Punuk sites are also associated with evidence of raiding and warfare in the form of slat armor, improved bow-and-arrow weaponry, and ivory daggers.[53]

At roughly the same time, smaller sites appear on both sides of the Bering Strait and spread eastward along the northern coast of Alaska (*Birnirk* phase). Although they comprise only a few houses, these sites contain most elements of the complex technology developed by the Old Bering Sea/Okvik

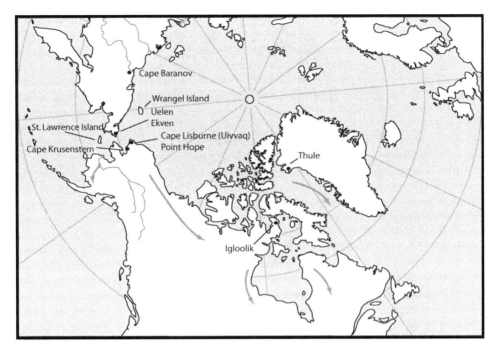

Figure 7.8. Map of sites of the Old Bering Sea/Okvik, Punuk/Birnirk, and Thule cultures of the circumpolar zone.

people. Whaling was clearly limited in comparison to later periods, but still may have been an important component of the Birnirk economy.[54]

By AD 500, a complex political geography had evolved around the Bering Strait region. Several polities were probably competing for control of the richest areas for sea-mammal hunting.[55] Among them, the most striking was *Ipiutak,* which had developed from a Norton base and was established on the Chukchi Sea coast of Alaska. The largest Ipiutak settlement was located at Point Hope and comprised hundreds of houses and more than a hundred burials. Despite its size—many of the houses were probably occupied at different times—settlement at Point Hope was seasonal. Like their Denbigh and Norton ancestors, Ipiutak people moved back and forth between the coast and the interior, pursuing a mixed economy.[56]

Inexplicably, they had abandoned key elements of Norton technology such as ground slate, lamps, and pottery. Their equipment for sea-mammal hunting was simpler than that of the Punuk/Birnirk people and apparently used primarily on seal. But though its potential for whaling was unexploited by the Ipiutak, Point Hope was nevertheless a sufficiently productive location to sustain a large settlement. In fact, variations among the burials suggest differences in social status. Some of the corpses were dismembered

Figure 7.9. Death mask of the Ipiutak culture from Point Hope.

before burial; others were dressed with bizarre death masks. The style of the art is often described as "Scytho-Siberian," and fantastic-looking seals, bears, and humanlike creatures are depicted. Many archaeologists believe that Ipiutak society was dominated by powerful shamans.[57]

Climates were generally warm during the first millennium, but around AD 900 a brief cold oscillation took place. For reasons that are unclear, Ipiutak vanished at this time and Point Hope was subsequently occupied by Thule or Inuit people, who remain there today. Climates warmed again within a hundred years as the Medieval Warm Period began (described in chapter 1). Reduced sea-ice cover improved conditions for open-water whaling, and the descendants of the Old Bering Sea/Okvik people—now classified as Thule—mobilized their sophisticated technology to harvest this immense resource on a unprecedented scale.[58]

At an average weight of 100 tons, a single bowhead whale *(Balaena mysticetus)* provided an enormous windfall of food, fuel, and material. Bowheads could be killed in sufficient numbers as they migrated north during the spring, supporting increasingly large village settlements. Hunting them on the open water in umiaks required whaling crews under the leadership of a captain, promoting an almost military organization.[59]

Thule settlements expanded rapidly after AD 1000 into southern Alaska and—more significantly—eastward into the summer feeding areas of the whales in the central Arctic. A trail of village sites (each composed of the remains of several houses) yielding Thule harpoon heads and large quantities of whalebone allows archaeologists to track the rapid expansion of the

Thule people across northern Canada. By AD 1300 they had reached northern Greenland.[60]

If Paleo-Eskimo peoples had sometimes abandoned proven technologies (such as lamps), the Thule widened their already extensive array of implements and facilities with more innovations. In addition to their large boats and whaling equipment, they developed a suite of implements for hunting ringed and bearded seals on the winter ice, including a trip alarm to indicate their arrival at a breathing hole. Polar foxes were caught in small box traps with sliding doors, and caribou were hunted with a sinew-backed bow. At some point, Thule people engineered the complex technology for dog traction, involving large sleds with runners, harness toggles, trace buckles, and other accessories.[61]

During the long winters, Thule families occupied spacious semi-subterranean houses. These were usually about 15 ft (5 m) in diameter and either rectangular or oval in plan. A sunken entrance tunnel prevented cold air from seeping into the living area, and the frame was insulated with hides and sod. The interior was warmed and illuminated with sea-mammal oil lamps. Family members slept comfortably on baleen mattresses over raised platforms. Games and toys helped pass the time.[62]

Figure 7.10. The small Thule (Inuit) village at Uivvaq near Cape Lisburne, Alaska, photographed in March 1922.

The Thule invaders seem to have completely overwhelmed the Dorset people, who disappeared within a short period of time. The process might have been similar to the spread of modern humans into the Neanderthal areas of northern Eurasia (see chapter 5). The ability of the Thule people to harvest a larger share of local resources with their superior food-getting technology may have been a critical factor. However, their weapons technology, whaling crew organization, and tradition of warfare probably accelerated the process. There is evidence—much of it derived from Inuit oral history—for violent conflict between the Thule and the Dorset.[63]

By AD 1450 climates had cooled again with the beginning of the Little Ice Age (also described in chapter 1). As usual, this comparatively minor climate change had a powerful impact on circumpolar environments and their human inhabitants. Increased sea ice prevented bowhead whales from reaching their summer feeding areas, and Thule people had abandoned the islands of the central High Arctic by the late 1600s. During this period, Thule groups also expanded southward into Labrador, where they eventually confronted European whalers.[64]

In many parts of the Arctic, Thule peoples responded to the Little Ice Age with significant changes in their economy. On Baffin Island, they shifted from whaling to walrus hunting and ceased construction of winter houses and umiaks. In the region west of Hudson Bay, they developed an economy based chiefly on interior resources—fish and caribou—and also abandoned many elements of their earlier technology. The variety of local responses to cooler climates contributed to the breakdown of the Thule way of life and the diversity of Inuit dialects and groups encountered by Europeans after AD 1500.[65]

Cold War in the Circumpolar Zone

Cold climate and geographic remoteness—combined with the required costs of coping with both—seem to have discouraged large-scale settlement of the Arctic by industrial civilizations of the temperate zone. At the beginning of the third millennium AD, not one major urban center could be found at or above 66° North (although several major cities may be found above 60° North in the Subarctic). The limited agricultural possibilities of the Arctic appear to be a lesser factor—major cities are found, after all, in desert areas where biotic productivity is much lower. Nor has a lack of other resources—from fur-bearing mammals and whales to oil—inhibited industrial expansion into the circumpolar zone.

As a result of the limited invasion of civilizations from the temperate zone, the native peoples of the Arctic have continued to maintain some degree of isolation from other cultures. This isolation is breaking down, however, primarily because of recent advances in industrial transportation and communication technology (including cable television and the Internet), which have intensified the impact of external cultures without large-scale settlement.

Global conflict in the mid-twentieth century AD—not population expansion or the lure of natural resources—brought the first significant intrusion into many areas of the Arctic. The circumstances were created by a conjunction of political geography—the juxtaposition of competing powers on opposite sides of the North Pole—and innovations in industrial military technology.

As international tensions increased following the end of World War II, the Soviet Union deployed long-range bombers at airfields in northeast Siberia that could reach targets in North America across the circumpolar zone. The United States responded by establishing a network of radar stations in Alaska. One of them was built at Cape Lisburne on the Chukchi Sea coast (latitude 68° North) at a location *(Uivvaq)* used since Birnirk/Thule times for hunting seal, walrus, and caribou. The American military struggled with the novel problems of constructing and operating a facility in an arctic environment. The radar and communications equipment functioned poorly at high latitudes.[66]

To strengthen its own strategic bomber threat, the United States also constructed a massive air base in 1951–1952 at the original Thule site in northwest Greenland—less than 2,500 miles (4,000 km) from Moscow. The U.S. Air Force was quickly adapting to arctic conditions and employed novel engineering technology in the construction of the facility. Barracks were constructed on stilt foundations with pipes for cold air circulation (to prevent

Figure 7.11. Thule Air Base in northwest Greenland.

thawing of frozen ground), insulated walls, and double arctic doors with plastic handles. The local Inuit community was relocated.[67]

In 1955 the United States began construction of the Distant Early Warning (DEW) system—a line of fifty-seven radar stations stretching across the North American Arctic from Cape Lisburne to Greenland. The radars were newly designed for high latitude operation, and the living quarters were prefabricated modules for cold-climate habitation. The DEW Line stations had a profound and lasting impact on the Canadian Inuit population, ending the subsistence economy.[68]

But by the time the DEW Line became operational in 1957, the Soviet Union had developed an intercontinental ballistic missile, which could not be detected by existing radars. A Soviet missile launch complex was established in a forward location near Plesetsk in the subarctic forest zone of north European Russia. Accordingly, the United States began construction of a missile early warning system—also focused on the polar regions.[69] This system remained operational throughout the Cold War, which ended in 1989–1991 with the collapse of the Soviet Union.

At Cape Lisburne, some of the original radar facility was pulled down in the summer of 2002, where the writing of this book began. The remains of the Birnirk/Thule houses of Uivvaq were still present.

Notes

Chapter 1. Vikings in the Arctic

1. Hubert H. Lamb, *Climate, History, and the Modern World*, 2nd ed. (London: Routledge, 1995), pp. 171–186; Brian M. Fagan, *The Little Ice Age: How Climate Made History, 1300–1850* (New York: Basic Books, 2000), pp. 3–21. There is some uncertainty about the geographic extent of the warming trend, which may have been less than global in scope.

2. Erik Wahlgren, *The Vikings and America* (London: Thames and Hudson, 1986); Peter Schledermann, "Ellesmere: Vikings in the Far North," in *Vikings: The North Atlantic Saga,* ed. W. W. Fitzhugh and E. I. Ward, pp. 248–256 (Washington, DC: Smithsonian Institution Press, 2000).

3. Patricia Sutherland, "The Norse and Native North Americans," in *Vikings,* ed. Fitzhugh and Ward, pp. 238–247. The term *skraeling* has been translated as "weakling" or "barbarian." Hans Christian Gulløv, "Natives and Norse in Greenland," in *Vikings,* ed. Fitzhugh and Ward, pp. 318–326; Wendell H. Oswalt, *Eskimos and Explorers*, 2nd ed. (Lincoln: University of Nebraska Press, 1999), p. 5.

4. Robert McGhee, *Ancient People of the Arctic* (Vancouver: University of British Columbia, 1996); Daniel Odess, Stephen Loring, and William W. Fitzhugh, "Skraeling: First Peoples of Helluland, Markland, and Vinland," in *Vikings,* ed. Fitzhugh and Ward, pp. 193–205.

5. Moreau S. Maxwell, *Prehistory of the Eastern Arctic* (Orlando, FL: Academic Press, 1985), pp. 247–294. Wendell Oswalt developed a simple method for quantifying the complexity of tools and weapons and applied it to various recent hunter-gatherer peoples, including the Inuit. See Wendell H. Oswalt, "Technological Complexity: The Polar Eskimos and the Tareumiut," *Arctic Anthropology* 24, no. 2 (1987): 82–98.

6. Oswalt, *Eskimos and Explorers,* pp. 5–24; Odess, Loring, and Fitzhugh, "Skraeling," pp.193–205.

7. Wahlgren, *The Vikings and America,* pp. 169–177; Lamb, *Climate, History, and the Modern World,* pp. 187–210.

8. Jette Arneborg et al., "Change of Diet of the Greenland Vikings Determined from Stable Carbon Isotope Analysis and 14c Dating of Their Bones," *Radiocarbon* 41, no. 2 (1999): 157–168; Niels Lynnerup, "Life and Death in Norse Greenland," in *Vikings,* ed. Fitzhugh and Ward, pp. 285–294.

9. Oswalt, *Eskimos and Explorers,* pp. 19–20; Thomas H. McGovern, "The Demise of Norse Greenland," in *Vikings,* ed. Fitzhugh and Ward, pp. 327–339. Although many of the Viking settlements in Greenland were situated at roughly the same latitude as Iceland (roughly 65° North), Greenland temperatures are several degrees lower than the latter because of the influence of ocean currents.

10. John F. Hoffecker and Scott A. Elias, "Environment and Archeology in Beringia," *Evolutionary Anthropology* 12, no. 1 (2003): 34–49.

11. N. K. Vereshchagin and G. F. Baryshnikov, "Paleoecology of the Mammoth Fauna in the Eurasian Arctic," in *Paleoecology of Beringia,* ed. D. M. Hopkins et al., pp. 267–279 (New York: Academic Press, 1982); McGhee, *Ancient People of the Arctic,* pp. 110–116.

12. As one moves away from the equator, the angle of sunlight increases and equal units of solar energy penetrate thicker layers of the atmosphere and fall on larger areas of the surface. Steven B. Young, *To the Arctic: An Introduction to the Far Northern World* (New York: John Wiley and Sons, 1994), pp. 5–12. At latitude 40° North, the average temperature is about 60° F (14° C), while at 50° North, the annual mean falls to 42° F (6° C). Eric R. Pianka, *Evolutionary Ecology,* 2nd ed. (New York: Harper and Row Publishers, 1978), pp. 18–25.

13. Plant productivity is typically measured in terms of the amount of organic matter created per unit area each year, and it declines with falling temperature and moisture. While a tropical rain forest produces an average of 2,200 grams of organic matter per square meter each year, arctic tundra environments may produce as little as 140 grams per year. Robert H. Whittaker, *Communities and Ecosystems,* 2nd ed. (New York: Macmillan Publishing Co., 1975), pp. 192–231.

14. Ann Henderson-Sellers and Peter J. Robinson, *Contemporary Climatology* (Edinburgh Gate: Addison Wesley Longman, 1986), pp. 79–85.

15. Clive Gamble, "The Earliest Occupation of Europe: The Environmental Background," in *The Earliest Occupation of Europe,* ed. W. Roebroeks and T. van Kolfschoten, pp. 279–295 (Leiden: University of Leiden, 1995), p. 283.

Chapter 2. Out of Africa

1. Russell H. Tuttle, *Apes of the World: Their Social Behavior, Communication, Mentality, and Ecology* (Park Ridge, NJ: Noyes Publications, 1986), pp. 12–20.

2. Daniel J. Boorstin, *The Discoverers: A History of Man's Search to Know His World and Himself* (New York: Random House, 1983), pp. 459–463; Ian Tattersall, *The Fossil Trail: How We Know What We Think We Know about Human Evolution* (New York: Oxford University Press, 1995), p. 4.

3. Tattersall, *The Fossil Trail,* p. 4.

4. Tuttle, *Apes of the World,* pp. 12–20, 171–196.

5. Ibid., pp. 40–52.

6. Charles Darwin, *The Descent of Man and Selection in Relation to Sex* (London: John Murray, 1871).

7. Jane Goodall, "My Life among the Wild Chimpanzees," *National Geographic Magazine* 124 (1963): 272–308. Since 1963 much new information about chimpanzee production and use of tools has been collected. William C. McGrew, *Chimpanzee Material Culture* (Cambridge: Cambridge University Press, 1992); Yukimaru Sugiyama, "Social Tradition and the Use of Tool-Composites by Wild Chimpanzees," *Evolutionary Anthropology* 6, no. 1 (1997): 23–27.

8. Geza Teleki, *The Predatory Behavior of Wild Chimpanzees* (Lewisburg, PA: Bucknell University Press, 1973); Tuttle, *Apes of the World*, pp. 63–113.

9. Lawrence Martin, "Significance of Enamel Thickness in Hominid Evolution," *Nature* 314 (1985): 260–263.

10. Tattersall, *The Fossil Trail*, pp. 122–123.

11. J. Marks, C. W. Schmid, and V. M. Sarich, "DNA Hybridization as a Guide to Phylogeny: Relations of the Hominoidea," *Journal of Human Evolution* 17 (1988): 769–786; W. Enard et al., "Intra- and Interspecific Variation in Primate Gene Expression Patterns," *Science* 296 (2002): 340–343; Elizabeth Pennisi, "Jumbled DNA Separates Chimps and Humans," *Science* 298 (2002): 719–721.

12. V. M. Sarich and A. C. Wilson, "Immunological Time Scale for Hominid Evolution," *Science* 158 (1967): 1200–1203.

13. For example, see David Pilbeam, *The Ascent of Man: An Introduction to Human Evolution* (New York: Macmillan Publishing Co. 1969), pp. 91–99.

14. Roger Lewin, *Bones of Contention: Controversies in the Search for Human Origins* (New York: Simon and Schuster, 1987), pp. 108–127.

15. Peter Andrews, "Hominoid Evolution," *Nature* 295 (1982): 185–186.

16. Russell H. Tuttle, "Knuckle-walking and the Problems of Human Origins," *Science* 166 (1969): 953–961; Martin, "Significance of Enamel Thickness."

17. Glenn C. Conroy, *Primate Evolution* (New York: W. W. Norton and Co., 1990), pp. 185–194; Richard G. Klein, *The Human Career*, 2nd ed. (Chicago: University of Chicago Press, 1999), pp. 126–127.

18. Klein, *The Human Career*, pp. 119–126.

19. Laszlo Kordos and David R. Begun, "Rudabanya: A Late Miocene Subtropical Swamp Deposit with Evidence of the Origin of the African Apes and Humans," *Evolutionary Anthropology* 11 (2002): 45–57.

20. Conroy, *Primate Evolution*, pp. 229–240; Klein, *The Human Career*, pp. 134–140.

21. Conroy, *Primate Evolution*, pp. 185–205.

22. Klein, *The Human Career*, pp. 186–187.

23. John Napier, "The Antiquity of Human Walking," *Scientific American* 216, no. 4 (1967): 56–66.

24. Peter S. Rodman and Henry M. McHenry, "Bioenergetics of Hominid Bipedalism," *American Journal of Physical Anthropology* 52 (1980): 103–106.

25. Roger Lewin, *Human Evolution: An Illustrated Introduction*, 3rd ed. (Boston: Blackwell Scientific Publications, 1993), pp. 85–90; Klein, *The Human Career*, pp. 249–250.

26. Lewin, *Human Evolution*, p. 85.

27. R. L. Susman, J. T. Stern, and W. L. Jungers, "Arboreality and Bipedality in the Hadar Hominids," *Folia Primatologica* 43 (1984): 113–156; F. B. Spoor, B. Wood, and F. Zonneveld, "Implications of Early Hominid Labyrinthine Morphology for Evolution of Human Bipedal Locomotion," *Nature* 369 (1994): 645–648.

28. John T. Robinson, "Adaptive Radiation in the Australopithecines and the Origin of Man," in *African Ecology and Human Evolution*, ed. F. C. Howell and

F. Bourliere, pp. 385–416 (Chicago: Aldine, 1963); Martin, "Significance of Enamel Thickness"; Lewin, *Human Evolution*, pp. 91–95.

29. Klein, *The Human Career*, pp. 186–187. Stable carbon isotope analysis of australopithecine teeth indicate heavy consumption of foods such as grasses or sedges (that is, foods enriched by carbon-13) or the animals that eat such foods. Matt Spoonheimer and Julia A. Lee-Thorp, "Isotopic Evidence for the Diet of an Early Hominid, *Australopithecus africanus*," *Science* 283 (1999): 368–370.

30. L.S.B. Leakey, P. V. Tobias, and J. R. Napier, "A New Species of the Genus *Homo* from Olduvai Gorge, Tanzania," *Nature* 202 (1964): 7–9; Lewin, *Bones of Contention*, pp. 142–151; B. Wood, "Origin and Evolution of the Genus *Homo*," *Nature* 355 (1982): 783–790.

31. Wood, "Origin and Evolution."

32. R. L. Susman and J. T. Stern, "Functional Morphology of *Homo habilis*," *Science* 217 (1982): 931–934.

33. Robinson, "Adaptive Radiation."

34. Glynn Ll. Isaac, "The Archaeology of Human Origins," *Advances in World Archaeology* 3 (1984): 1–87.

35. Mary D. Leakey, *Olduvai Gorge: Excavations in Beds I and II, 1960–1963* (Cambridge: Cambridge University Press, 1971); Nicholas Toth, "The Oldowan Reassessed: A Close Look at Early Stone Artifacts," *Journal of Archaeological Science* 12 (1985): 101–120. Analysis of microscopic damage, which is visible on a small percentage of Oldowan tools, indicates that the latter were used to cut meat, plants, and wood. Lawrence H. Keeley and Nicolas Toth, "Microwear Polishes on Early Stone Tools from Koobi Fora, Kenya," *Nature* 293 (1981): 464–465.

36. Kathy D. Schick and Nicholas Toth, *Making Silent Stones Speak: Human Evolution and the Dawn of Technology* (New York: Simon and Schuster, 1993), pp. 135–140.

37. R. Potts and P. Shipman, "Cutmarks Made by Stone Tools on Bones from Olduvai Gorge, Tanzania," *Nature* 291 (1981): 577–580; Pat Shipman, "Scavenging or Hunting in Early Hominids," *American Anthropologist* 88 (1986): 27–43.

38. Alan Walker and Mark Teaford, "Inferences from Quantitative Analysis of Dental Microwear," *Folia Primatologica* 53 (1989): 177–189.

39. Glynn Ll. Isaac, "The Food-Sharing Behavior of Protohuman Hominids," *Scientific American* 238, no. 4 (1978): 90–108.

40. Lewin, *Human Evolution*, pp. 135–140; Klein, *The Human Career*, pp. 239–248.

41. Clive Gamble, *Timewalkers: The Prehistory of Global Colonization* (Cambridge: Harvard University Press, 1994), pp. 125–134; Leo Gabunia et al., "Dmanisi and Dispersal," *Evolutionary Anthropology* 10 (2001): 158–170; R. X. Zhu et al., "Earliest Presence of Humans in Northeast Asia," *Nature* 413 (2001): 413–417.

42. Plant productivity in most areas of Africa above latitude 16° North—and Eurasia above 30° North—is less than half that of equatorial Africa. See O. W. Archibold, *Ecology of World Vegetation* (London: Chapman and Hall, 1995).

43. Susan Cachel and J.W.K. Harris, "The Lifeways of *Homo erectus* Inferred from Archaeology and Evolutionary Ecology: A Perspective from East Africa," in *Early*

Human Behaviour in Global Context: The Rise and Diversity of the Lower Palaeolithic Record, ed. M. D. Petraglia and R. Korisetter, pp. 108–132 (London: Routledge, 1998), pp. 113–123; Klein, *The Human Career,* p. 316.

44. M. Sahnouni and J. de Heinzelin, "The Site of Ain Hanech Revisited: New Investigations at This Lower Pleistocene Site in Northern Algeria," *Journal of Archaeological Science* 25 (1998): 1083–1101; Leo Gabunia et al., "Earliest Pleistocene Cranial Remains from Dmanisi, Republic of Georgia: Taxonomy, Geological Setting, and Age," *Science* 288 (2000): 1019–1025; Gabunia et al., "Dmanisi and Dispersal," pp. 162–164.

45. For example, see Alan Walker and Pat Shipman, *The Wisdom of the Bones: In Search of Human Origins* (New York: Alfred A. Knopf, 1996), pp. 240–241.

46. Gabunia et al., "Dmanisi and Dispersal."

47. Schick and Toth, *Making Silent Stones Speak,* pp. 254–257; Zhu et al., "Earliest Presence of Humans in Northeast Asia."

48. Carl C. Swisher, Garniss H. Curtis, and Roger Lewin, *Java Man* (New York: Scribner, 2000); J. de Vos, P. Sondaar, and C. C. Swisher, "Dating Hominid Sites in Indonesia," *Science* 266 (1994): 1726–27.

49. Abesalom Vekua et al., "A New Skull of Early *Homo* from Dmanisi, Georgia," *Science* 297 (2002): 85–89.

50. Klein, *The Human Career,* pp. 280–295.

51. Walker and Shipman, *The Wisdom of the Bones,* pp. 178–201.

52. Klein, *The Human Career,* pp. 289–291.

53. Christopher Ruff, "Climate, Body Size, and Body Shape in Human Evolution," *Journal of Human Evolution* 21 (1991): 81–105; Walker and Shipman, *The Wisdom of the Bones,* pp. 195–199.

54. Robert G. Franciscus and Erik Trinkaus, "Nasal Morphology and the Emergence of *Homo erectus,*" *American Journal of Physical Anthropology* 75, no. 4 (1988): 517–527.

55. Klein, *The Human Career,* pp. 292–293.

56. Gamble, *Timewalkers,* pp. 141–143.

57. Klein, *The Human Career,* p. 181; Ofer Bar-Yosef and Anna Belfer-Cohen, "From Africa to Eurasia—Early Dispersals," *Quaternary International* 75 (2001): 19–28. Although generally rare in East Asia, hand axes dating to 0.8 million years ago were recently reported from the Bose Basin in southern China. Yamei Hou et al., "Mid-Pleistocene Acheulean-like Stone Technology of the Bose Basin, South China," *Science* 287 (2000): 1622–26.

58. Schick and Toth, *Making Silent Stones Speak,* pp. 258–260.

59. Ibid., pp. 237–245; Thomas G. Wynn, "The Evolution of Tools and Symbolic Behaviour," in *Handbook of Human Symbolic Evolution,* ed. Andrew Lock and Charles R. Peters, pp. 263–287 (Oxford: Blackwell Publishers, 1999), pp. 268–271.

60. Thomas Wynn, "Handaxe Enigmas," *World Archaeology* 27, no. 1 (1995): 10–24.

61. C. K. Brain and A. Sillent, "Evidence from the Swartkrans Cave for the Earliest Use of Fire," *Nature* 336 (1988): 464–466; Klein, *The Human Career,* pp. 350–351. As in the case of hand axes, the controlled use of fire may have important

implications for human cognition. See Derek Bickerton, *Language and Species* (Chicago: University of Chicago Press, 1990), pp. 140–141. For many years, the earliest confirmed use of fire was attributed to the cave of Zhoukoudian in China, which was occupied roughly 400,000 years ago. However, a recent analysis of the deposits at Zhoukoudian Locality 1 found no evidence for fire. P.Goldberg et al., "Site Formation Processes at Zhoukoudian, China," *Journal of Human Evolution* 41 (2001): 483–530.

62. Walker and Shipman, *The Wisdom of the Bones*, pp. 168–169.

63. C. M. Monahan, "New Zooarchaeological Data from Bed II, Olduvai Gorge, Tanzania: Implications for Hominid Behavior in the Early Pleistocene," *Journal of Human Evolution* 31 (1996): 93–128; Cachel and Harris, "The Lifeways of *Homo erectus*," pp. 114–115.

64. Robert L. Kelly, *The Foraging Spectrum: Diversity in Hunter-Gatherer Lifeways* (Washington, DC: Smithsonian Institution Press, 1995), pp. 66–73.

Chapter 3. The First Europeans

1. Wil Roebroeks and Thijs van Kolfschoten, "The Earliest Occupation of Europe: A Reappraisal of Artefactual and Chronological Evidence," in *The Earliest Occupation of Europe*, ed. W. Roebroeks and T. van Kolfschoten, pp. 297–315 (Leiden: University of Leiden, 1995); Richard G. Klein, *The Human Career*, 2nd ed. (Chicago: University of Chicago Press, 1999), pp. 319–327.

2. W. von Koenigswald, "Various Aspects of Migrations in Terrestrial Animals in Relation to Pleistocene Faunas of Central Europe." *Courier Forschungsinstitut Senckenberg* 153 (1992): 39–47; Clive Gamble, "The Earliest Occupation of Europe: The Environmental Background," in *The Earliest Occupation of Europe*, ed. Roebroeks and van Kolfschoten, pp. 279–295. For example, on the Atlantic coast of Norway at latitude 69° North, Tromso experiences a mean annual temperature of 37° F (2° C) and a January mean of 25° F ($-3°$ C). But in northeast Siberia at a comparable latitude (67° North), Verkhoyansk records a mean annual temperature of only 7° F ($-13°$ C) and a January mean of $-48°$ F ($-44°$ C).

3. Gamble, "The Earliest Occupation of Europe."

4. Clive Gamble, *Timewalkers: The Prehistory of Global Colonization* (Cambridge: Harvard University Press, 1994), pp. 135–136.

5. Alan Turner, "Large Carnivores and Earliest European Hominids: Changing Determinants of Resource Availability during the Lower and Middle Pleistocene," *Journal of Human Evolution* 22 (1992): 109–126.

6. Neanderthal adaptations to cold climate are described in chapter 4. At a minimum, they included various anatomical features that reduced body heat loss and a very high protein and fat diet.

7. Wil Roebroeks, Nicholas J. Conard, and Thijs van Kolfschoten, "Dense Forests, Cold Steppes, and the Palaeolithic Settlement of Northern Europe," *Current Anthropology* 33 (1992): 551–586; Gamble, "The Earliest Occupation of Europe," p. 281.

8. Roebroeks and van Kolfschoten, "The Earliest Occupation of Europe," pp. 307–308.

9. John F. Hoffecker, *Desolate Landscapes: Ice-Age Settlement in Eastern Europe* (New Brunswick, NJ: Rutgers University Press, 2002), pp. 40–42.

10. Roebroeks and van Kolfschoten, "The Earliest Occupation of Europe," pp. 303–308.

11. Ibid., pp. 496–499; Jean-Paul Raynal, Lionel Magoga, and Peter Bindon, "Tephrofacts and the First Human Occupation of the French Massif Central," in *The Earliest Occupation of Europe*, ed. Roebroeks and van Kolfschoten, pp. 129–146; Karel Valoch, "The Earliest Occupation of Europe: Eastern Central and Southeastern Europe," in *The Earliest Occupation of Europe*, ed. Roebroeks and van Kolfschoten, pp. 67–84.

12. Eudald Carbonell and Xose Pedro Rodriguez, "Early Middle Pleistocene Deposits and Artefacts in the Gran Dolina site (TD4) of the 'Sierra de Atapuerca' (Burgos, Spain)," *Journal of Human Evolution* 26 (1994): 291–311. TD Level 6 reportedly underlies the Brunhes/Matuyama paleomagnetic boundary, which is currently dated at 780,000 years ago. J. M. Pares and A. Peres-Gonzalez, "Magnetochronology and Stratigraphy at Gran Dolina Section, Atapuerca (Burgos, Spain)," *Journal of Human Evolution* 37 (1999): 325–342.

13. J. M. Bermudez de Castro et al., "A Hominid from the Lower Pleistocene of Atapuerca, Spain: Possible Ancestor to Neandertals and Modern Humans," *Science* 276 (1997): 1392–95. The paleoanthropologist G. Philip Rightmire has observed that the separate species classification *Homo antecessor* is based on a small sample that is both fragmentary and includes subadult specimens, and that the Atapuerca people may turn out to be part of the same group that includes later European hominids such as *Homo heidelbergensis*. See G. Philip Rightmire, "Patterns of Hominid Evolution and Dispersal in the Middle Pleistocene," *Quaternary International* 75 (2001): 77–84.

14. Carbonell and Rodriguez, "Early Middle Pleistocene Deposits," pp. 301–305; E. Carbonell et al., "Lower Pleistocene Hominids and Artifacts from Atapuera-TD6 (Spain)," *Science* 269 (1995): 826–830; E. Carbonell et al., "The TD6 Level Lithic Industry from Gran Dolina, Atapuerca (Burgos, Spain): Production and Use," *Journal of Human Evolution* 37, nos. 3–4 (1999): 653–694.

15. Carbonell and Rodriguez, "Early Middle Pleistocene Deposits," pp. 300–301; Carbonell et al., "Lower Pleistocene Hominids and Artifacts," p. 826; J. Carlos Diez et al., "Zooarchaeology and Taphonomy of Aurora Stratum (Gran Dolina, Sierra de Atapuerca, Spain)," *Journal of Human Evolution* 37, nos. 3–4 (1999): 623–652; Yolanda Fernandez-Jalvo et al., "Human Cannibalism in the Early Pleistocene of Europe (Gran Dolina, Sierra de Atapuerca, Burgos, Spain," *Journal of Human Evolution* 37, nos. 3–4 (1999): 591–622.

16. Derek Roe, "The Orce Basin (Andalucia, Spain) and the Initial Palaeolithic of Europe," *Oxford Journal of Archaeology* 14 (1995): 1–12; Robin Dennell and Wil Roebroeks, "The Earliest Colonization of Europe: The Short Chronology Revisited," *Antiquity* 70 (1996): 535–542.

17. The estimated brain volume of the Ceprano hominid is 1,185 cc. A. Ascenzi et al., "A Calvarium of Late *Homo erectus* from Ceprano, Italy," *Journal of Human Evolution* 31 (1996): 409–423.

18. Roebroeks and van Kolfschoten, "The Earliest Occupation of Europe."

19. Hoffecker, *Desolate Landscapes,* pp. 36–40.

20. Ofer Bar-Yosef, "Pleistocene Connexions between Africa and Southwest Asia: An Archaeological Perspective," *African Archaeological Review* 5 (1987): 29–38; Idit Saragusti and Naama Goren-Inbar, "The Biface Assemblage from Gesher Benot Ya'aqov, Israel: Illuminating Patterns in 'Out of Africa' Dispersal," *Quaternary International* 75 (2001): 85–89. The earliest-known appearance of the hand ax makers in Europe may be in southern Italy at Venosa Notarchirico, which is dated to 640,000 years ago. See Paola Villa, "Early Italy and the Colonization of Western Europe," *Quaternary International* 75 (2001): 113–130.

21. G. Philip Rightmire, "Human Evolution in the Middle Pleistocene: The Role of *Homo heidelbergensis,*" *Evolutionary Anthropology* 6, no. 6 (1998): 218–227.

22. M. B. Roberts and S. A. Parfitt, eds., *Boxgrove: A Middle Pleistocene Hominid Site at Eartham Quarry, Boxgrove, West Sussex* (English Heritage Archaeological Report no. 17, 1999).

23. Reinhart Kraatz, "A Review of Recent Research on Heidelberg Man, *Homo erectus heidelbergensis,*" in *Ancestors: The Hard Evidence,* ed. E. Delson, pp. 268–271 (New York: Alan R. Liss, 1985); Gerhard Bosinski, "The Earliest Occupation of Europe: Western Central Europe," in *The Earliest Occupation of Europe,* ed. Roebroeks and van Kolfschoten, pp. 103–128.

24. Rightmire, "Human Evolution in the Middle Pleistocene," pp. 222–224.

25. G. Philip Rightmire, *The Evolution of* Homo erectus: *Comparative Anatomical Studies of an Extinct Human Species* (Cambridge: Cambridge University Press, 1990), pp. 204–233. For a summary of human skeletal remains from this time range, see Klein, *The Human Career,* pp. 268–269, table 5.3.

26. The tendency toward increased body size among northern representatives of warm-blooded animals is referred to as the "Bergmann rule," while the pattern of shortened extremities is known as the "Allen rule." Eric R. Pianka, *Evolutionary Ecology,* 2nd ed. (New York: Harper and Row Publishers, 1978), pp. 307–308. For a review of how these "rules" have been applied to modern humans in northern environments, see G. Richard Scott et al., "Physical Anthropology of the Arctic," in *The Arctic: Environment, People, Policy,* ed. M. Nuttall and T. V. Callaghan, pp. 339–373 (Amsterdam: Harwood Academic Publishers, 2000).

27. Ralph L. Holloway, "The Poor Brain of *Homo sapiens neanderthalensis:* See What You Please . . . ," in *Ancestors: The Hard Evidence,* ed. Delson, pp. 319–324.

28. M. B. Roberts, C. B. Stringer, and S. A. Parfitt, "A Hominid Tibia from Middle Pleistocene Sediments at Boxgrove, UK," *Nature* 369 (1994): 311–313; C. B. Stringer and E. Trinkaus, "The Human Tibia from Boxgrove," in *Boxgrove: A Middle Pleistocene Hominid Site at Eartham Quarry, Boxgrove, West Sussex,* ed. M. B. Roberts and S. A. Parfitt, pp. 420–422 (English Heritage Archaeological Report no. 17, 1999).

29. Stringer and Trinkans, "The Human Tibia from Boxgrove," p. 422; D. F. Roberts, "Body Weight, Race, and Climate," *American Journal of Physical Anthropology* 11 (1953): 533–558.

30. Turner, "Large Carnivores and Earliest European Hominids," pp. 113–121.

31. Forty-six percent of the 150 horse bones from Unit 4b and 25 percent of the 53 red deer bones from Unit 4c exhibit damage from stone tools at Boxgrove. S. A. Parfitt and M. B. Roberts, "Human Modification of Faunal Remains," in *Boxgrove*, ed. Roberts and Parfitt, pp. 395–415.

32. P. Anconetani, "Lo Studio Arcezoologico del Sito di Isernia La Pineta," in *I Reperti Paleontologici del Giacimento Paleolitico di Isernia La Pineta: L'uomo e L'ambiente*, ed. C. Peretto, pp. 87–186 (Isernia, 1996); A. Tuffreau et al., "Le Gisement Acheuléen de Cagny-L'Epinette (Somme)," *Bulletin de la Société Préhistorique Française* 92 (1995): 169–199; Hartmut Thieme, "Lower Palaeolithic Hunting Spears from Germany," *Nature* 385 (1997): 807–810.

33. Elaine Turner, "The Problems of Interpreting Hominid Subsistence Strategies at Lower Palaeolithic Sites—a Case Study from the Central Rhineland of Germany," in *Hominid Evolution: Lifestyles and Survival Strategies*, ed. H. Ullrich, pp. 365–382 (Gelsenkirchen-Schwelm: Edition Archaea, 1999); John F. Hoffecker, G. F. Baryshnikov, and V. B. Doronichev, "Large Mammal Taphonomy of the Middle Pleistocene Hominid Occupation at Treugol'naya Cave (Northern Caucasus)," *Quaternary Science Reviews* 22, nos. 5–7 (2003): 595–607.

34. P. F. Puech, A. Prone, and R. Kraatz, "Microscopie de l'Usure Dentaire chez l'Homme Fossile: Bol Alimentaire et Environnement," *CRASP* 290 (1980): 1413–1416.

35. G. Russell Coope, "Late-Glacial (Anglian) and Late-Temperate (Hoxnian) Coleoptera," in *The Lower Paleolithic Site at Hoxne, England*, ed. R. Singer, B. G. Gladfelter, and J. J. Wymer, pp. 156–162 (Chicago: University of Chicago Press, 1993).

36. Leslie G. Freeman, "Acheulean Sites and Stratigraphy in Iberia and the Maghreb," in *After the Australopithecines*, ed. K. W. Butzer and G. Ll. Isaac, pp. 661–743 (The Hague: Mouton Publishers, 1975), pp. 664–682; Lewis R. Binford, *Bones: Ancient Men and Modern Myths* (New York: Academic Press, 1981); Sabine Gaudzinski and Elaine Turner, "The Role of Early Humans in the Accumulation of European Lower and Middle Palaeolithic Bone Assemblages," *Current Anthropology* 37 (1996): 153–156.

37. P. Shipman and J. Rose, "Evidence of Butchery and Hominid Activities at Torralba and Ambrona: An Evaluation Using Microscopic Techniques," *Journal of Archaeological Science* 10 (1983): 465–474; Richard G. Klein, "Problems and Prospects in Understanding How Early People Exploited Animals," in *The Evolution of Human Hunting*, ed. M. H. Nitecki and D. V. Nitecki, pp. 11–45 (New York: Plenum Press, 1987).

38. Robert J. Blumenschine, "Early Hominid Scavenging Opportunities," *British Archaeological Reports International Series* 283 (1986).

39. M. Kretzoi and V. Dobosi, eds., *Vértesszöllös: Man, Site, and Culture* (Budapest: Akademiai Kiado, 1990); Ronald Singer, Bruce G. Gladfelter, and John J. Wymer,

The Lower Paleolithic Site at Hoxne, England (Chicago: University of Chicago Press, 1993).

40. Two of the only cave occupations known from this period provide conflicting information. At Arago in the French Pyrenees, reindeer remains seem to have been collected by humans (see Turner, "The Problems of Interpreting Hominid Subsistence Strategies," p. 380) and suggest central-place foraging, while at Treugol'naya Cave in the northern Caucasus, the large mammal remains may have accumulated by natural processes. See Hoffecker, Baryshnikov, and Doronichev, "Large Mammal Taphonomy."

41. Glyn Daniel, *The Idea of Prehistory* (Harmondsworth: Penguin Books, 1962), pp. 32–46.

42. Kathy D. Schick and Nicholas Toth, *Making Silent Stones Speak: Human Evolution and the Dawn of Technology* (New York: Simon and Schuster, 1993), pp. 231–245; Thomas G. Wynn, "The Evolution of Tools and Symbolic Behaviour," in *Handbook of Human Symbolic Evolution,* ed. Andrew Lock and Charles R. Peters, pp. 263–287 (Oxford: Blackwell Publishers, 1999).

43. Roberts and Parfitt, *Boxgrove.*

44. Lawrence H. Keeley, "Microwear Analysis of Lithics," in *The Lower Paleolithic Site at Hoxne, England,* ed. Singer, Gladfelter, and Wymer, pp. 129–138.

45. Ibid.; Lawrence H. Keeley, *Experimental Determination of Stone Tool Uses: A Microwear Analysis* (Chicago: University of Chicago Press, 1980), pp. 86–119; John J. Wymer and Ronald Singer, "Flint Industries and Human Activity," in *The Lower Paleolithic Site at Hoxne, England,* ed. Singer, Gladfelter, and Wymer, pp. 74–128.

46. Schick and Toth, *Making Silent Stones Speak,* pp. 270–271; Thieme, "Lower Palaeolithic Hunting Spears from Germany." Although the wood artifact from Clacton-on-Sea was originally reported to exhibit traces of burning—providing some evidence for the use of fire—more recent analyses have determined that this was incorrect. See Steven R. James, "Hominid Use of Fire in the Lower and Middle Pleistocene," *Current Anthropology* 30, no. 1 (1989): 1–26.

47. Gamble, *Timewalkers,* p. 138; Klein, *The Human Career,* pp. 349–350.

48. James, "Hominid Use of Fire," pp. 6–9; P. Villa and F. Bon, "Fire and Fireplaces in the Lower, Middle, and Early Upper Paleolithic of Western Europe," *Journal of Human Evolution* 42, no. 3 (2002): A37–A38.

49. Clive Gamble, *The Palaeolithic Settlement of Europe* (Cambridge: Cambridge University Press, 1986), pp. 387–390.

50. Hallam L. Movius, "The Lower Paleolithic Cultures of Southern and Eastern Asia," *Transactions of the American Philosophical Society* 38 (1948): 329–420; Schick and Toth, *Making Silent Stones Speak,* pp. 276–279.

51. Leo Gabunia et al., "Dmanisi and Dispersal," *Evolutionary Anthropology* 10 (2001): 158–170; Ofer Bar-Yosef and Anna Belfer-Cohen, "From Africa to Eurasia—Early Dispersals," *Quaternary International* 75 (2001): 19–28; Carl C. Swisher, Garniss H. Curtis, and Roger Lewin, *Java Man* (New York: Scribner, 2000).

52. Charles B. M. McBurney, "The Geographical Study of the Older Palaeolithic Stages in Europe," *Proceedings of the Prehistoric Society* 16 (1950): 163–183; Kretzoi and Dobosi, *Vértesszöllös;* Dietrich Mania, "The Earliest Occupation of Eu-

rope: The Elbe-Saale Region (Germany)," in *The Earliest Occupation of Europe*, ed. Roebroeks and van Kolfschoten, pp. 85–101.

53. Emanuel Vlček, "Patterns of Human Evolution," in *Hunters between East and West: The Paleolithic of Moravia*, by J. Svoboda, V. Ložek, and E. Vlček, pp. 37–74 (New York: Plenum Press, 1996), pp. 38–46; Klein, *The Human Career*, pp. 339–341.

54. Gamble, *The Palaeolithic Settlement of Europe*, pp. 141–146; Paola Villa, *Terra Amata and the Middle Pleistocene Archaeological Record of Southern France* (Berkeley: University of California Press, 1983). West European archaeologists traditionally assigned the hand ax and non–hand ax sites to different industries. Sites containing large bifaces were placed in the Acheulean, while those lacking hand axes were often assigned to the "Clactonian" or other Lower Paleolithic industries of Europe.

55. McBurney, "Geographical Study."

56. Kretzoi and Dobosi, *Vértesszöllös*; Mania, "The Earliest Occupation of Europe," pp. 91–92; Thieme, "Lower Palaeolithic Hunting Spears from Germany."

57. Rightmire, "Human Evolution in the Middle Pleistocene."

58. Turner, "Large Carnivores and Earliest European Hominids."

59. Major glacial episodes during this period are thought to have occurred at 480,000–430,000 years ago and 350,000–340,000 years ago. For example, see J. J. Lowe and M.J.C. Walker, *Reconstructing Quaternary Environments*, 2nd ed. (London: Longman, 1997).

60. Roebroeks, Conard, and van Kolfschoten, "Dense Forests, Cold Steppes"; Gamble, "The Earliest Occupation of Europe," p. 281; Mania, "The Earliest Occupation of Europe," p. 97; Alain Tuffreau and Pierre Antoine, "The Earliest Occupation of Europe: Continental Northwestern Europe," in *The Earliest Occupation of Europe*, ed. Roebroeks and van Kolfschoten, pp. 147–163.

Chapter 4. Cold Weather People

1. John F. Hoffecker, *Desolate Landscapes: Ice-Age Settlement in Eastern Europe* (New Brunswick, NJ: Rutgers University Press, 2002), pp. 55–62.

2. Erik Trinkaus and Pat Shipman, *The Neandertals: Of Skeletons, Scientists, and Scandal* (New York: Vintage Books, 1994), pp. 3–90. Neanderthal skulls had been discovered as early as 1830 in Belgium (and also in 1848 in Spain), but were not recognized as early human remains until many years after the Neander Valley discovery.

3. Christopher Stringer and Clive Gamble, *In Search of the Neanderthals: Solving the Puzzle of Human Origins* (New York: Thames and Hudson, 1993), pp. 13–15; Trinkaus and Shipman, *The Neandertals*.

4. Stringer and Gamble, *In Search of the Neanderthals*, pp. 34–38; Ian Tattersall, *The Last Neanderthal: The Rise, Success, and Mysterious Extinction of Our Closest Human Relatives*, rev. ed. (Boulder, CO: Westview Press, 1999), pp. 111–116.

5. Stringer and Gamble, *In Search of the Neanderthals*, pp. 26–33; Trinkaus and Shipman, *The Neandertals*.

6. Niles Eldredge, *Time Frames: The Evolution of Punctuated Equilibria* (Princeton: Princeton University Press, 1985).

7. M. Krings et al., "Neanderthal DNA Sequences and the Origin of Modern Humans," *Cell* 90 (1997): 19–30; M. Krings et al., "A View of Neandertal Genetic Diversity," *Nature Genetics* 26, no. 2 (2000): 144–146; I. V. Ovchinnikov et al., "Molecular Analysis of Neanderthal DNA from the Northern Caucasus," *Nature* 404 (2000): 490–493.

8. G. Philip Rightmire, "Human Evolution in the Middle Pleistocene: The Role of *Homo heidelbergensis*," *Evolutionary Anthropology* 6, no. 6 (1998): 218–227; Richard G. Klein, *The Human Career*, 2nd ed. (Chicago: University of Chicago Press, 1999), pp. 295–312.

9. J.-J. Hublin, "Climatic Changes, Paleogeography, and the Evolution of the Neandertals," in *Neandertals and Modern Humans in Western Asia*, ed. T. Akazawa, K. Aoki, and O. Bar-Yosef, pp. 295–310 (New York: Plenum Press, 1998).

10. C. B. Stringer, "Secrets of the Pit of the Bones," *Nature* 362 (1993): 501–502.

11. Stringer and Gamble, *In Search of the Neanderthals*, pp. 65–69.

12. Hublin, "Climatic Changes," pp. 300–301.

13. Stringer and Gamble, *In Search of the Neanderthals*, pp. 74–84.

14. Ibid., pp. 86–95.

15. Paul Mellars, *The Neanderthal Legacy: An Archaeological Perspective from Western Europe* (Princeton: Princeton University Press, 1996), pp. 56–140. One of the earliest Mousterian sites is Maastricht-Belvédère in the Netherlands, which is dated to the interglacial period between 245,000 and 185,000 years ago. See Wil Roebroeks, "Archaeology and Middle Pleistocene Stratigraphy: The Case of Maastricht-Belvédère (NL)," in *Chronostratigraphie et Faciès Culturels du Paléolithique Inférieur et Moyen dans l'Europe de Nord-Ouest*, ed. A. Tuffreau and J. Somme, pp. 81–86 (Paris: Supplément au Bulletin de l'Association Française pour l'Étude du Quaternaire, 1986).

16. Carleton S. Coon, *The Origin of Races* (New York: Alfred A. Knopf, 1962), pp. 529–547; T. W. Holliday, "Postcranial Evidence of Cold Adaptation in European Neandertals," *American Journal of Physical Anthropology* 104 (1997): 245–258.

17. Ralph L. Holloway, "The Poor Brain of *Homo sapiens neanderthalensis*: See What You Please...," in *Ancestors: The Hard Evidence*, ed. Delson, pp. 319–324.

18. Trinkaus and Shipman, *The Neandertals*, pp. 316–324.

19. Coon, *The Origin of Races*, pp. 546–548; Erik Trinkaus, "Neanderthal Limb Proportions and Cold Adaptation," in *Aspects of Human Evolution*, ed. C. Stringer, pp. 187–224 (London: Taylor and Francis, 1981); Holliday, "Postcranial Evidence."

20. Coon, *The Origin of Races*, p. 543; Robert G. Franciscus and Steven E. Churchill, "The Costal Skeleton of Shanidar 3 and a Reappraisal of Neandertal Thoracic Morphology," *Journal of Human Evolution* 42 (2002): 303–356.

21. Coon, *The Origin of Races*, p. 534.

22. Ibid., pp. 532–534; Erik Trinkaus, "Bodies, Brawn, Brains, and Noses: Human Ancestors and Human Predation," in *The Evolution of Human Hunting*, ed. M. Nitecki and D. V. Nitecki, pp. 107–145 (New York: Plenum Press, 1987); Trinkaus and Shipman, *The Neandertals*, pp. 317–321, p. 417.

23. Coon, *The Origin of Races*, pp. 541–542; C. Loring Brace, "The Fate of the 'Classic' Neanderthals: A Consideration of Hominid Catastrophism," *Current An-*

thropology 5 (1964): 3–43; W. L. Hylander, "The Adaptive Significance of Eskimo Cranio-Facial Morphology," in *Oro-Facial Growth and Development*, ed. A. A. Dahlberg and T. Graber (The Hague: Mouton, 1977); Erik Trinkaus, *The Shanidar Neandertals* (New York: Academic Press, 1983).

24. Trinkaus, "Bodies, Brawn, Brains and Noses: Human Ancestors and Human Predation"; Klein, *The Human Career*, pp. 388–389.

25. Henri Laville, Jean-Philippe Rigaud, and James Sackett, *Rock Shelters of the Perigord* (New York: Academic Press, 1980), pp. 179–215; Mellars, *The Neanderthal Legacy*, pp. 32–55.

26. Laville, Rigaud, and Sackett, *Rock Shelters of the Perigord*, p. 189; Clive Gamble, *The Palaeolithic Settlement of Europe* (Cambridge: Cambridge University Press, 1986), pp. 367–369.

27. Laville, Rigaud, and Sackett, *Rock Shelters of the Perigord*, pp. 197–201; Hublin, "Climatic Changes," p. 305.

28. Ovchinnikov et al., "Molecular Analysis of Neanderthal DNA."

29. Jiří Svoboda, Vojen Ložek, and Emanuel Vlček, *Hunters between East and West: The Paleolithic of Moravia* (New York: Plenum Press, 1996), pp. 82–85; Hublin, "Climatic Changes," p. 306; Hoffecker, *Desolate Landscapes*, pp. 64–65. Mousterian sites in Eastern Europe antedating the Last Interglacial may include *Korolevo* (Danubian Basin) and *Khryashchi* and *Mikhailovskoe* on the Severskii Donets River.

30. A. A. Velichko, "Late Pleistocene Spatial Paleoclimatic Reconstructions," in *Late Quaternary Environments of the Soviet Union*, ed. A. A. Velichko, pp. 261–285 (Minneapolis: University of Minnesota Press, 1984), pp. 261–273.

31. F. M. Zavernyaev, *Khotylevskoe Paleoliticheskoe Mestonakhozhdenie* (Leningrad: Nauka, 1978); Hoffecker, *Desolate Landscapes*, pp. 72–74.

32. Hoffecker, *Desolate Landscapes*, pp. 65–66. A similar pattern of abandonment during the Lower Pleniglacial, followed by reoccupation during the milder interstadial that began after 60,000 years ago, is also evident in parts of Central Europe. See Gamble, *The Palaeolithic Settlement of Europe*, pp. 374–377.

33. Z. A. Abramova, "Must'erskii Grot Dvuglazka v Khakasii (Predvaritel'noe Soobshschenie)," *Kratkie Soobshcheniya Instituta Arkheologii* 165 (1981): 74–78. Stone artifacts from the open-air site of *Diring Yuriakh* overlooking the Lena River at latitude 61° North—investigated by Yuri Mochanov—are dated by TL to more than 260,000 years ago. See Michael R. Waters, Steven L. Forman, and James M. Pierson, "Diring Yuriakh: A Lower Paleolithic Site in Central Siberia." *Science* 275 (1997): 1281-1284. This site might represent an early intrusion into high latitudes in Siberia (although some archaeologists believe that the artifacts are naturally fractured rocks), but appears unrelated to the Neanderthals. Diring is more likely tied to Lower Paleolithic peoples in the Far East, who possibly moved north during a warm interglacial of the later Middle Pleistocene (described in chapter 3).

34. Ted Goebel, "The Pleistocene Colonization of Siberia and Peopling of the Americas: An Ecological Approach," *Evolutionary Anthropology* 8 (1999): 208–227.

35. Ofer Bar-Yosef, "Upper Pleistocene Cultural Stratigraphy in Southwest

Asia," in *The Emergence of Modern Humans*, ed. E. Trinkaus, pp. 154–180 (Cambridge: Cambridge University Press, 1989); Hublin, "Climatic Changes," pp. 305–307.

36. Trinkaus, "Neanderthal Limb Proportions and Cold Adaptation," p. 215; Franciscus and Churchill, "The Costal Skeleton of Shanidar 3," pp. 352–353. Also perhaps significant is recent evidence for plant food consumption at Amud Cave in Israel. See Marco Madella et al., "The Exploitation of Plant Resources by Neanderthals in Amud Cave (Israel): The Evidence from Phytolith Studies," *Journal of Archaeological Science* 29 (2002): 703–719.

37. Coon, *The Origin of Races*, p. 534.

38. Kathy D. Schick and Nicholas Toth, *Making Silent Stones Speak: Human Evolution and the Dawn of Technology* (New York: Simon and Schuster, 1993), p. 292; Klein, *The Human Career*, p. 328.

39. Mellars, *The Neanderthal Legacy*, pp. 56–94.

40. Thomas Wynn, "Piaget, Stone Tools and the Evolution of Human Intelligence," *World Archaeology* 17, no. 1 (1985): 32–43.

41. A description of composite tools and weapons among recent nonindustrial peoples is provided in Wendell H. Oswalt, *An Anthropological Analysis of Food-Getting Technology* (New York: John Wiley and Sons, 1976).

42. Sylvie Beyries, "Functional Variability of Lithic Sets in the Middle Paleolithic," in *Upper Pleistocene Prehistory of Western Eurasia*, ed. H. L. Dibble and A. Montet-White, pp. 213–224 (Philadelphia: University of Pennsylvania Museum, 1988), pp. 219–220; John J. Shea, "Spear Points from the Middle Paleolithic of the Levant," *Journal of Field Archaeology* 15 (1988): 441–450; Patricia Anderson-Gerfaud, "Aspects of Behaviour in the Middle Palaeolithic: Functional Analysis of Stone Tools from Southwest France," in *The Emergence of Modern Humans*, ed. P. Mellars, pp. 389–418 (Edinburgh: Edinburgh University Press, 1990), pp. 402–410; E. Boëda et al., "Bitumen as a Hafting Material on Middle Paleolithic Artefacts," *Nature* 380 (1996): 336–338.

43. Nicholas Rolland and Harold L. Dibble, "A New Synthesis of Middle Paleolithic Variability," *American Antiquity* 55 (1990): 480–499; Mellars, *The Neanderthal Legacy*, pp. 95–140.

44. Hallam L. Movius, "A Wooden Spear of Third Interglacial Age from Lower Saxony," *Southwestern Journal of Anthropology* 6 (1950): 139–142; E. Carbonell and Z. Castro-Curel, "Palaeolithic Wooden Artifacts from the Abric Romani (Capellades, Barcelona Spain)," *Journal of Archaeological Science* 19 (1992): 707–719; Z. Castro-Curel and E. Carbonell, "Wood Pseudomorphs from Level I at Abric Romani, Barcelona, Spain," *Journal of Field Archaeology* 22 (1995): 376–384.

45. Beyries, "Functional Variability of Lithic Sets," pp. 214–219; Anderson-Gerfaud, "Aspects of Behaviour in the Middle Palaeolithic," pp. 401–404.

46. Wendell H. Oswalt, "Technological Complexity: The Polar Eskimos and the Tareumiut," *Arctic Anthropology* 24, no. 2 (1987): 82–98.

47. N. D. Praslov, "Paleolithic Cultures in the Late Pleistocene," in *Late Quaternary Environments of the Soviet Union*, ed. A. A. Velichko, pp. 313–318 (Minneapolis: University of Minnesota Press, 1984), p. 314; Anderson-Gerfaud, "Aspects of Behaviour in the Middle Palaeolithic," p. 405.

48. Hoffecker, *Desolate Landscapes*, pp. 106–107.

49. Ibid., pp. 107–108. The most frequently cited example of a possible artificial shelter was recorded on the floor of Layer 4 at *Molodova I* in the Dnestr Valley (southwest Ukraine), but few archaeologists believe that this was actually an enclosed structure. See Richard G. Klein, *Ice-Age Hunters of the Ukraine* (Chicago: University of Chicago Press, 1973), pp. 69–70.

50. *Quest for Fire* was based on a 1911 short story entitled "La Guerre du Feu" by J. H. Rosny-Aine. See Stringer and Gamble, *In Search of the Neanderthals*, pp. 30–31. Recent hunter-gatherer peoples who reportedly lacked the ability to produce fire include the Andaman Islanders, Mbuti (Pygmies), and aboriginal Tasmanians. See Carleton S. Coon, *The Hunting Peoples* (New York: Little, Brown and Co., 1971).

51. Mellars, *The Neanderthal Legacy*, pp. 295–301; Hoffecker, *Desolate Landscapes*, pp. 108–109.

52. H. Bocherens et al., "Palaeoenvironmental and Palaeodietary Implications of Isotopic Biogeochemistry of Last Interglacial Neanderthal and Mammal Bones at Scladina Cave (Belgium)," *Journal of Archaeological Science* 26 (1999): 599–607; M. P. Richards et al., "Neanderthal Diet at Vindija and Neanderthal Predation: The Evidence from Stable Isotopes," *Proceedings of the National Academy of Sciences* 97, no. 13 (2000): 7663–66.

53. Coon, *The Origin of Races*, p. 534.

54. Philip G. Chase, "The Hunters of Combe Grenal: Approaches to Middle Paleolithic Subsistence in Europe," *British Archaeological Reports International Series* S-286 (1986); Gennady Baryshnikov and John F. Hoffecker, "Mousterian Hunters of the NW Caucasus: Preliminary Results of Recent Investigations," *Journal of Field Archaeology* 21 (1994): 1–14.

55. Chase, "The Hunters of Combe Grenal"; Sabine Gaudzinski, "On Bovid Assemblages and Their Consequences for the Knowledge of Subsistence Patterns in the Middle Palaeolithic," *Proceedings of the Prehistoric Society* 62 (1996): 19–39; Curtis W. Marean and Zelalem Assefa, "Zooarchaeological Evidence for the Faunal Exploitation Behavior of Neandertals and Early Modern Humans," *Evolutionary Anthropology* 8, no. 1 (1999): 22–37; John F. Hoffecker and Naomi Cleghorn, "Mousterian Hunting Patterns in the Northwestern Caucasus and the Ecology of the Neanderthals," *International Journal of Osteoarchaeology* 10 (2000): 368–378. Some evidence for scavenging of large mammals is reported from sites in west-central Italy. See Mary C. Stiner, *Honor among Thieves: A Zooarchaeological Study of Neandertal Ecology* (Princeton: Princeton University Press, 1994).

56. Baryshnikov and Hoffecker, "Mousterian Hunters of the NW Caucasus," pp. 10–11; Hoffecker, *Desolate Landscapes*, pp. 113–115. At La Cotte de St. Brelade mammoth remains exhibit tool marks and reflect predominance of prime-age adults. See Katherine Scott, "Mammoth Bones Modified by Humans: Evidence from La Cotte de St. Brelade, Jersey, Channel Islands," in *Bone Modification*, ed. R. Bonnichsen and M. H. Sorg, pp. 335–346 (Orono, ME: Center for the Study of the First Americans, 1989).

57. Stiner, *Honor among Thieves*, pp. 158–198; Hoffecker and Cleghorn, "Mousterian Hunting Patterns."

58. M. C. Stiner et al., "Paleolithic Population Growth Pulses Evidenced by Small Animal Exploitation," *Science* 283 (1999): 190–194. Most recent foraging peoples in areas above latitude 50° North consume a diet composed of at least 20 per cent fish. See Robert L. Kelly, *The Foraging Spectrum: Diversity in Hunter-Gatherer Lifeways* (Washington, DC: Smithsonian Institution Press, 1995), pp. 66–73.

59. Baryshnikov and Hoffecker, "Mousterian Hunters of the NW Caucasus"; L. V. Golovanova et al., "Mezmaiskaya Cave: A Neanderthal Occupation in the Northern Caucasus," *Current Anthropology* 41 (1999): 77–86; Hoffecker and Cleghorn, "Mousterian Hunting Patterns."

60. Edwin N. Wilmsen, "Interaction, Spacing Behavior, and the Organization of Hunting Bands," *Journal of Anthropological Research* 29 (1973): 1–31; Hoffecker, *Desolate Landscapes*, p. 112.

61. J. T. Laitman, R. C. Heimbuch, and C. S. Crelin, "The Basicranium of Fossil Hominids as an Indicator of Their Upper Respiratory Systems," *American Journal of Physical Anthropology* 51 (1979): 15–34; P. Lieberman et al., "The Anatomy, Physiology, Acoustics, and Perception of Speech: Essential Elements in the Analysis of the Evolution of Human Speech," *Journal of Human Evolution* 23 (1992): 447–467.

62. Stringer and Gamble, *In Search of the Neanderthals*, pp. 88–91.

63. Iain Davidson and William Noble, "The Archaeology of Perception: Traces of Depiction and Language," *Current Anthropology* 30, no. 2 (1989): 125–155.

64. Philip G. Chase and Harold L. Dibble, "Middle Paleolithic Symbolism: A Review of Current Evidence and Interpretations," *Journal of Anthropological Archaeology* 6 (1987): 263–296; Mellars, *The Neanderthal Legacy*, pp. 369–375.

65. Francesco d'Errico et al., "A Middle Palaeolithic Origin of Music? Using Cave-Bear Bone Accumulations to Assess the Divje Babe I Bone 'Flute,'" *Antiquity* 72 (1998): 65–79.

66. Francis B. Harrold, "A Comparative Analysis of Eurasian Palaeolithic Burials," *World Archaeology* 12, no. 2 (1980): 195–211; Trinkaus and Shipman, *The Neandertals*. The complete absence of burials in Neanderthal open-air sites suggests that lack of graves in earlier time periods (prior to 250,000 years ago) might simply reflect the scarcity of caves and rock shelters from these periods.

67. The most famous "grave goods" reported from a Neanderthal burial are the flowers—preserved in the form of concentrations of pollen—at Shanidar Cave in Iraq (Burial No. 4) described by Ralph S. Solecki, *Shanidar, the First Flower People* (New York: Alfred Knopf, 1971). It appears likely that the pollen concentrations are derived from flower heads collected by rodents that burrowed into the cave sediment in and around this burial. See Jeffrey D. Sommer, "The Shanidar IV 'Flower Burial': A Re-evaluation of Neanderthal Burial Ritual," *Cambridge Archaeological Journal* 9, no. 1 (1999): 127–129.

68. Chase and Dibble, "Middle Paleolithic Symbolism," pp. 273–274; Stringer and Gamble, *In Search of the Neanderthals*, pp. 158–160; Mellars, *The Neanderthal Legacy*, pp. 375–381.

69. Clive Gamble, *Timewalkers: The Prehistory of Global Colonization* (Cambridge: Harvard University Press, 1994), pp. 167–174.

70. Lewis R. Binford, "Hard Evidence," *Discover* February (1992): 44–51; Mellars, *The Neanderthal Legacy*, pp. 357–359.

71. Kelly, *The Foraging Spectrum*, pp. 262–270; Mellars, *The Neanderthal Legacy*, pp. 361–362.

72. Mellars, *The Neanderthal Legacy*, pp. 270–295; Hoffecker, *Desolate Landscapes*, p. 131.

73. Kelly, *The Foraging Spectrum*, pp. 210–213.

74. Wil Roebroeks, J. Kolen, and E. Rensink, "Planning Depth, Anticipation and the Organization of Middle Palaeolithic Technology: the 'Archaic Natives' Meet Eve's Descendents," *Helinium* 28, no. 1 (1988): 17–34; J. Feblot-Augustins, "Raw Material Transport Patterns and Settlement Systems in the European Lower and Middle Palaeolithic: Continuity, Change, and Variability," in *The Middle Palaeolithic Occupation of Europe*, ed. W. Roebroeks and C. Gamble, pp. 193–214 (Leiden: University of Leiden, 1999).

75. Stringer and Gamble, *In Search of the Neanderthals*, pp. 195–218; Tattersall, *The Last Neanderthal*, pp. 198–203.

76. Coon, *The Origin of Races*.

77. Hoffecker, *Desolate Landscapes*, pp. 157–158.

78. Milford H. Wolpoff, *Paleoanthropology*, 2nd ed. (Boston: McGraw-Hill, 1999).

79. C. Duarte et al., "The Early Upper Paleolithic Human Skeleton from the Abrigo do Lagar Velho (Portugal) and Modern Human Emergence in Iberia," *Proceedings of the National Academy of Sciences* 96 (1999): 7604–7609. A modern human jaw recovered from Romania and dated to more than 35,000 years ago is said to possess at least one Neanderthal trait (unilateral mandibular foramen lingular bridging). See Erik Trinkaus et al., "An Early Modern Human from the Pestera cu Oase, Romania," *Proceedings of the National Academy of Sciences* 100, no. 20 (2003): 11231–36.

80. Richard G. Klein, "Whither the Neanderthals?" *Science* 299 (2003): 1525–27.

Chapter 5. Modern Humans in the North

1. Clive Gamble, *Timewalkers: The Prehistory of Global Colonization* (Cambridge: Harvard University Press, 1994), pp. 181–202; Ted Goebel, "The Pleistocene Colonization of Siberia and Peopling of the Americas: An Ecological Approach," *Evolutionary Anthropology* 8 (1999): 208–227. Calendar years—not radiocarbon years before present—are used in this chapter and throughout the book. Past variations in atmospheric radiocarbon distort carbon-14 dates by as much as several thousand years, but these distortions can be corrected by a calibration scale. See, for example, C. Bronk-Ramsey, "Radiocarbon Calibration and Analysis of Stratigraphy: The OxCal Program," *Radiocarbon* 37 (1995): 425–430. For a new high-resolution calibration scale for the past 50,000 years, see K. Hughen et al., "^{14}C Activity and Global Carbon Cycle Changes over the Past 50,000 Years." *Science* 303 (2004): 202–207.

2. Richard G. Klein, *The Human Career*, 2nd ed. (Chicago: University of Chicago Press, 1999), pp 520–544.

3. John F. Hoffecker, *Desolate Landscapes: Ice-Age Settlement in Eastern Europe* (New Brunswick, NJ: Rutgers University Press, 2002), pp. 158–162.

4. Paul Mellars, "Major Issues in the Emergence of Modern Humans." *Current Anthropology* 30, no. 3 (1989): 349–385; Christopher Stringer and Robin McKie, *African Exodus: The Origins of Modern Humanity* (New York: Henry Holt and Co., 1996), pp. 194–223.

5. Goebel, "The Pleistocene Colonization of Siberia," p. 218; P. Dolukhanov, D. Sokoloff, and A. Shukurov, "Radiocarbon Chronology of Upper Palaeolithic Sites in Eastern Europe at Improved Resolution," *Journal of Archaeological Science* 28 (2001): 699–712; Hoffecker, *Desolate Landscapes*, pp. 200–201.

6. John F. Hoffecker and Scott A. Elias, "Environment and Archeology in Beringia," *Evolutionary Anthropology* 12, no. 1 (2003): 34–49.

7. Klein, *The Human Career*, pp. 305–312.

8. For example, see F. L. Coolidge and T. Wynn, "Executive Functions of the Frontal Lobes and the Evolutionary Ascendancy of *Homo sapiens*," *Journal of Human Evolution* 42, no. 3 (2002): A12–A13. For general background on the subject, see Elkhonon Goldberg, *The Executive Brain: Frontal Lobes and the Civilized Mind* (Oxford: Oxford University Press, 2001).

9. Philip Lieberman, *Eve Spoke: Human Language and Human Evolution* (New York: W. W. Norton and Co., 1998), pp. 85–96.

10. G. E. Kennedy, "The Emergence of *Homo sapiens*: The Post-cranial Evidence," *Man* 19 (1984): 94–110.

11. Derek Bickerton, *Language and Species* (Chicago: University of Chicago Press, 1990), pp. 75–104.

12. Ibid., pp. 7–24. Ironically, one of the only known examples of animal communication of mental models is found not among mammals or other vertebrates, but in the insect world. The famous honeybee dance—used to communicate the location of a food source—is an isolated case of this type of communication. See Derek Bickerton, *Language and Human Behavior* (Seattle: University of Washington Press, 1995), pp. 12–18.

13. S. McBrearty and A. S. Brooks, "The Revolution That Wasn't: A New Interpretation of the Origin of Modern Human Behavior," *Journal of Human Evolution* 39, no. 5 (2000): 453–563.

14. Klein, *The Human Career*, pp. 514–517.

15. Bickerton suggests that *Homo erectus* and other premodern humans probably evolved a "proto-language" without syntax and similar to the speech of modern human children under the age of two years. See Bickerton, *Language and Species*, pp. 164–197. The evolution of the capacity for syntax—not merely using symbols—may be the most important difference between modern humans and other animals. William H. Calvin and Derek Bickerton, *Lingua ex Machina: Reconciling Darwin and Chomsky with the Human Brain* (Cambridge: MIT Press, 2000), pp. 19–24.

16. Wolfgang Enard et al., "Molecular Evolution of FOXP2, a Gene Involved in Speech and Language," *Nature* 418 (2002): 869–872.

17. Klein, *The Human Career*, pp. 439–440; Richard G. Klein, "Archaeology and the Evolution of Human Behavior," *Evolutionary Anthropology* 9 (2000): 17–36.

18. Christopher S. Henshilwood et al., "An Early Bone Tool Industry from the Middle Stone Age at Blombos Cave, South Africa: Implications for the Origins of Modern Human Behaviour, Symbolism, and Language," *Journal of Human Evolution* 41, no. 6 (2001): 631–678.

19. Brian M. Fagan, *The Journey from Eden: The Peopling of Our World* (London: Thames and Hudson, 1990); Gamble, *Timewalkers*, pp. 181–202. Some vertebrates, such as the foxes and canids, exhibit very broad longitudinal and latitudinal distribution but have differentiated into distinct species (or sometimes different genera) within their range.

20. Goebel, "The Pleistocene Colonization of Siberia," pp. 213–216; J. F. Hoffecker et al., "Initial Upper Paleolithic in Eastern Europe: New Research at Kostenki," *Journal of Human Evolution* 42, no. 3 (2002): A16–A17.

21. R. E. Taylor, "Radiocarbon Dating: The Continuing Revolution," *Evolutionary Anthropology* 4 (1996): 169–181.

22. James K. Feathers, "Luminescence Dating and Modern Human Origins," *Evolutionary Anthropology* 5, no. 1 (1996): 25–36. Much of the chronology for the transition to modern humans in the Near East now rests on thermoluminescence (TL) dates, supported by some ESR dating. See Ofer Bar-Yosef, "The Middle and Upper Paleolithic in Southwest Asia and Neighboring Regions," in *The Geography of Neandertals and Modern Humans in Europe and the Greater Mediterranean*, ed. O. Bar-Yosef and D. Pilbeam, pp. 107–156 (Cambridge, MA: Peabody Museum of Archaeology and Ethnology, 2000).

23. Christopher Stringer and Clive Gamble, *In Search of the Neanderthals: Solving the Puzzle of Human Origins* (New York: Thames and Hudson, 1993), pp. 96–122.

24. Ofer Bar-Yosef, "Upper Pleistocene Cultural Stratigraphy in Southwest Asia," in *The Emergence of Modern Humans*, ed. E. Trinkaus, pp. 154–180 (Cambridge: Cambridge University Press, 1989).

25. Lieberman, *Eve Spoke*, pp. 137–143. The early modern humans in the Levant ca. 100,000 years ago did practice intentional burial like their Neanderthal counterparts in Europe and the Near East. See also Bar-Yosef, "The Middle and Upper Paleolithic," p. 119.

26. M. M. Lahr and R. Foley, "Multiple Dispersals and Modern Human Origins," *Evolutionary Anthropology* 3 (1994): 48–60; R. G. Roberts et al., "The Human Colonisation of Australia: Optical Dates of 53,000 and 60,000 Years Bracket Human Arrival at Deaf Adder Gorge, Northern Territory," *Quaternary Science Reviews* 13 (1994): 575–586; J. M. Bowler et al., "New Ages for Human Occupation and Climatic Change at Lake Mungo, Australia," *Nature* 421 (2003): 837–840.

27. Guanjun Shen et al., "U-Series Dating of Liujiang Hominid Site in Guangxi, Southern China," *Journal of Human Evolution* 43 (2002): 817–829.

28. Lahr and Foley, "Multiple Dispersals."

29. Bar-Yosef, "The Middle and Upper Paleolithic," pp. 111–130.

30. Hoffecker et al., "Initial Upper Paleolithic in Eastern Europe"; A. A. Sinitsyn, "Nizhnie Kul'turnye Sloi Kostenok 14 (Markina Gora) (Raskopki 1998–2001 gg.)," in *Kostenki v Kontekste Paleolita Evrazii,* ed. A. A. Sinitsyn, V. Ya. Sergin, and J. F. Hoffecker, pp. 219–236 (St. Petersburg: Russian Academy of Sciences, 2002); M. V. Anikovich, "The Early Upper Paleolithic in Eastern Europe." *Archaeology, Ethnology & Anthropology of Eurasia* 2, no. 14 (2003): 15–29.

31. Ted Goebel, A. P. Derevianko, and V. T. Petrin, "Dating the Middle-to-Upper-Paleolithic Transition at Kara-Bom," *Current Anthropology* 34 (1993): 452–458; Goebel, "The Pleistocene Colonization of Siberia," pp. 213–214.

32. V. P. Alekseev, "The Physical Specificities of Paleolithic Hominids in Siberia," in *The Paleolithic of Siberia: New Discoveries and Interpretations,* ed. A. P. Derevianko, pp. 329–335 (Urbana: University of Illinois Press, 1998); Goebel, "The Pleistocene Colonization of Siberia," pp. 215–218.

33. Paul Mellars, *The Neanderthal Legacy: An Archaeological Perspective from Western Europe* (Princeton: Princeton University Press, 1996), pp. 392–419; Jean-Pierre Bocquet-Appel and Pierre Yves Demars, "Neanderthal Contraction and Modern Human Colonization of Europe," *Antiquity* 74 (2000): 544–552.

34. Stringer and Gamble, *In Search of the Neanderthals,* pp. 179–181; Klein, *The Human Career,* pp. 496–497. In 2003 a modern human jaw was reported from a cave in Romania with a radiocarbon date of 36,000–34,000 years before present. See Erik Trinkaus et al., "An Early Modern Human from the Pestera cu Oase, Romania," *Proceedings of the National Academy of Sciences* 100, no. 20 (2003): 11231–36.

35. Hoffecker, *Desolate Landscapes,* p. 143.

36. Erik Trinkaus, "Neanderthal Limb Proportions and Cold Adaptation," in *Aspects of Human Evolution,* ed. C. Stringer, pp. 187–224 (London: Taylor and Francis, 1981); T. W. Holliday, "Brachial and Crural Indices of European Late Upper Paleolithic and Mesolithic Humans," *Journal of Human Evolution* 36 (1999): 549–566; Hoffecker, *Desolate Landscapes,* pp. 155–158.

37. A. A. Velichko et al., "Periglacial Landscapes of the East European Plain," in *Late Quaternary Environments of the Soviet Union,* ed. A. A. Velichko, pp. 94–118 (Minneapolis: University of Minnesota Press, 1984), p. 114.

38. Pavel Pavlov, John Inge Svendsen, and Svein Indrelid, "Human Presence in the European Arctic nearly 40,000 Years Ago," *Nature* 413 (2001): 64–67; V. V. Pitulko et al., "The Yana RHS Site: Humans in the Arctic before the Last Glacial Maximum," *Science* 303 (2004): 52–56.

39. David Lewis-Williams, *The Mind in the Cave: Consciousness and the Origins of Art* (London: Thames and Hudson, 2002), pp. 29–36.

40. For example, John M. Lindly and Geoffrey A. Clark, "Symbolism and Modern Human Origins," *Current Anthropology* 31 (1990): 233–261.

41. Especially significant are the paintings of Chauvet Cave in southern France, which have been directly dated to 32,410–30,340 radiocarbon years before present. See Jean-Marie Chauvet, Eliette Brunel Deschamps, and Christian

Hillaire, *Dawn of Art: The Chauvet Cave* (New York: Harry N. Abrams, 1996), pp. 121–126. For recent discoveries of portable art of comparable age from Germany, see Nicholas J. Conard, "Palaeolithic Ivory Sculptures from Southwestern Germany and the Origins of Figurative Art." *Nature* 426 (2003): 830–832.

42. Sinitsyn, "Nizhnie Kul'turnye Sloi Kostenok 14," pp. 228–230.

43. Iain Davidson and William Noble, "The Archaeology of Perception: Traces of Depiction and Language," *Current Anthropology* 30, no. 2 (1989): 125–155.

44. Claude Lévi-Strauss, *The Savage Mind* (Chicago: University of Chicago Press, 1966), pp. 22–30.

45. D. Buisson, "Les Flûtes Paléolithiques d'Isturitz (Pyrénées Atlantiques)," *Société Préhistorique Française* 87 (1991): 420–433; Joachim Hahn, "Le Paléolithique Supérieur en Allemagne Méridonale (1991–1995)," *ERAUL* 76 (1996): 181–186.

46. S. N. Bibikov, *Drevneishii muzykal'nyi kompleks iz kostei mamonta* (Kiev: Naukova dumka, 1981).

47. The study of music among recent nonindustrial peoples of the world provides some insight into the variety of instruments that might have been made and used 35,000 years ago. See John E. Kaemmer, *Music in Human Life: Anthropological Perspectives on Music* (Austin: University of Texas Press, 1993), pp. 88–97; Jeff Todd Titon et al., *Worlds of Music: An Introduction to the Music of the World's Peoples* (New York: Schirmer Books, 1984).

48. Dean Falk, "Hominid Brain Evolution and the Origins of Music," in *The Origins of Music*, ed. N. L. Wallin, B. Merker, and S. Brown, pp. 197–216 (Cambridge: MIT Press, 2000).

49. Bruno Nettl, *Music in Primitive Culture* (Cambridge: Harvard University Press, 1956), pp. 73–74; Alan P. Merriam, *The Anthropology of Music* (Evanston, IL: Northwestern University Press, 1964), pp. 229–258; Kaemmer, *Music in Human Life*, pp. 108–141.

50. Clive Gamble, *The Palaeolithic Settlement of Europe* (Cambridge: Cambridge University Press, 1986), pp. 256–268. For background on how modern humans structure space ("proxemics") and time in different cultural contexts, see Edward T. Hall, *The Hidden Dimension* (Garden City, NY: Doubleday and Co., 1966) and Edward T. Hall, *The Dance of Life: The Other Dimension of Time* (New York: Doubleday and Co., 1983).

51. Alexander Marshack, *The Roots of Civilization* (London: Weidenfeld and Nicolson, 1972); Alexander Marshack, "The Taï Plaque and Calendrical Notation in the Upper Palaeolithic," *Cambridge Archaeological Journal* 1 (1991): 25–61; Karl W. Butzer et al., "Dating and Context of Rock Engravings in Southern Africa," *Science* 203 (1979): 1201–1214.

52. Francesco d'Errico, "Palaeolithic Lunar Calendars: a Case of Wishful Thinking?" *Current Anthropology* 30, no. 1 (1989): 117–118.

53. Paul G. Bahn and Jean Vertut, *Journey through the Ice Age* (Berkeley: University of California Press, 1997).

54. O. N. Bader, "Pogrebeniya v Verkhnem Paleolite i Mogila na Stoyanke Sungir'," *Sovetskaya Arkheologiya* 3 (1967): 142–159; O. N. Bader, "Vtoraya Paleo-

liticheskaya Mogila na Sungire," in *Arkheologicheskie Otkrytiya 1969 Goda,* pp. 41–43 (Moscow: Nauka, 1970); O. N. Bader, *Sungir' Verkhnepaleoliticheskaya Stoyanka* (Moscow: Nauka, 1978).

55. The application of black charcoal to the floor and red ochre to the filling of the graves suggests that these colors figured prominently in the burial ritual and beliefs associated with it. A comparative example among living people may be seen in the significance and role of the colors white, red, and black in the rituals of the Ndembu of Zambia. See Victor Turner, *The Forest of Symbols: Aspects of Ndembu Ritual* (Ithaca, NY: Cornell University Press, 1967).

56. The concept of "culture" has been defined by most of the major figures in anthropology. See, for example, A. L. Kroeber, *Anthropology* (New York: Harcourt, Brace and World, 1948); Leslie A. White, *The Science of Culture: A Study of Man and Civilization* (New York: Grove Press, 1949); Marvin Harris, *The Rise of Anthropological Theory: A History of Theories of Culture* (New York: Thomas Y. Crowell Co., 1968); Marshall Sahlins, *Culture and Practical Reason* (Chicago: University of Chicago Press, 1976).

57. Mary Douglas, "Symbolic Orders in the Use of Domestic Space," in *Man, Settlement, and Urbanism,* ed. P. J. Ucko, R. Tringham, and G. W. Dimbleby, pp. 513–521 (Cambridge, MA: Schenkman Publishing Co., 1970), p. 514.

58. M. Ingman et al., "Mitochondrial Genome Variation and the Origin of Modern Humans," *Nature* 408 (2000): 708–713.

59. Harold C. Conklin, "Lexicographical Treatment of Folk Taxonomies." *International Journal of American Linguistics* 28 (1962); Jared M. Diamond, "Zoological Classification System of a Primitive People," *Science* 151 (1966): 1102–1104; Lévi-Strauss, *The Savage Mind,* pp. 3–9.

60. V. Gordon Childe, *Man Makes Himself* (London: Watts and Co., 1936), p. 71; Bahn and Vertut, *Journey through the Ice Age,* pp. 134–158.

61. Gamble, *Timewalkers,* pp. 181–202.

62. Robert Whallon, "Elements of Cultural Change in the Later Paleolithic," in *The Human Revolution,* ed. P. Mellars and C. Stringer, pp. 433–454 (Princeton: Princeton University Press, 1989).

63. Sinitsyn, "Nizhnie Kul'turnye Sloi Kostenok 14," pp. 227–230.

64. Wil Roebroeks, J. Kolen, and E. Rensink, "Planning Depth, Anticipation, and the Organization of Middle Palaeolithic Technology: The 'Archaic Natives' Meet Eve's Descendents," *Helinium* 28, no. 1 (1988): 17–34; Clive Gamble, *The Palaeolithic Societies of Europe* (Cambridge: Cambridge University Press, 1999), pp. 313–319. Long-distance transport of materials is not evident among the earliest modern human sites in Siberia, but is documented after about 30,000 years ago. See Goebel, "The Pleistocene Colonization of Siberia and Peopling of the Americas," pp. 214–218.

65. Polly Wiessner, "Risk, Reciprocity and Social Influences on !Kung San Economics," in *Politics and History in Band Societies,* ed. E. Leacock and R. B. Lee, pp. 61–84 (Cambridge: Cambridge University Press, 1982); Bruce Winterhalder, "Diet Choice, Risk, and Food Sharing in a Stochastic Environment," *Journal of Anthropological Archaeology* 5 (1986): 369–392; Robert L. Kelly, *The Foraging Spectrum:*

Diversity in Hunter-Gatherer Lifeways (Washington, DC: Smithsonian Institution Press, 1995)

66. Gamble, *The Palaeolithic Settlement of Europe*, pp. 322–331.
67. Sahlins, *Culture and Practical Reason*.
68. Lévi-Strauss, *The Savage Mind*, pp. 8–9.
69. Bahn and Vertut, *Journey through the Ice Age*, pp. 171–183.
70. Gordon W. Hewes, "A History of Speculation on the Relation between Tools and Language," in *Tools, Language, and Cognition in Human Evolution*, ed. K. R. Gibson and T. Ingold, pp. 20–31 (Cambridge: Cambridge University Press, 1993); James Deetz, *Invitation to Archaeology* (Garden City, NY: Natural History Press, 1967), p. 86; Glynn Ll. Isaac, "Stages of Cultural Elaboration in the Pleistocene: Possible Archaeological Indicators of the Development of Language Capabilities," *Annals of the New York Academy of Sciences* 280 (1976): 275–288.

71. For discussion of the comparison between language and tool behavior, see Thomas G. Wynn, "The Evolution of Tools and Symbolic Behaviour," in *Handbook of Human Symbolic Evolution*, ed. Andrew Lock and Charles R. Peters, pp. 263–287 (Oxford: Blackwell Publishers, 1999), pp. 269–271.

72. For examples of the mathematical complexity of music and art among nonindustrial peoples, see Marc Chemillier, "Ethnomusicology, Ethnomathematics: The Logic Underlying Orally Transmitted Artistic Practices," in *Mathematics and Music*, ed. G. Assayag, H. G. Feichtinger, and J. F. Rodrigues, pp. 161–183 (Berlin: Springer-Verlag, 2002). For the structural analysis of myth, see, for example, Claude Lévi-Strauss, *Structural Anthropology*, trans. C. Jacobson and B. Schoepf (New York: Basic Books, 1963), pp. 206–231. Only the structural complexity of visual art is observable in the archaeological record; the complexity of music and myth is inferred on the basis of recent and living tribal peoples.

73. Kathy D. Schick and Nicholas Toth, *Making Silent Stones Speak: Human Evolution and the Dawn of Technology* (New York: Simon and Schuster, 1993), p. 355. A *mechanical* tool or device has been defined as one comprising two or more parts that alter their relationship to each other during its use. See Wendell H. Oswalt, *An Anthropological Analysis of Food-Getting Technology* (New York: John Wiley and Sons, 1976), p. 50.

74. Leslie White wrote, "It was the introduction of symbols . . . into the tool process that transformed anthropoid tool-behavior into human tool-behavior." See Leslie White, "On the Use of Tools by Primates," *Journal of Comparative Psychology* 34 (1942): 369–374. With respect to modern humans, it would be appropriate to substitute "syntax" for "symbols" (see earlier discussion in text and notes). The accelerated pace of innovation is apparent in John Troeng, *Worldwide Chronology of Fifty-three Prehistoric Innovations* (*Acta Archaeologica Lundensia* 8, no. 21, 1993). The aboriginal Tasmanians appear to be an example of people who spoke a fully modern language but produced a Lower Paleolithic technology lacking hafted tools, fire-making devices, and artificial shelters. However, they apparently had abandoned elements of more complex technology—a phenomenon observed among other peoples—after their isolation from mainland Australia, and still retained some forms of technology probably made only by modern humans (for example,

boats, sewn clothing, baited bird blinds). See Henry Lee Roth, *The Aborigines of Tasmania* (London: Kegan Paul, Trench, Trubner, 1890); Oswalt, *An Anthropological Analysis of Food-Getting Technology*, pp. 263–264.

75. Bruce G. Trigger, *A History of Archaeological Thought* (Cambridge: Cambridge University Press, 1989), pp. 94–102; Paul G. Bahn, ed., *The Cambridge Illustrated History of Archaeology* (Cambridge: Cambridge University Press, 1996), pp. 118–127.

76. François Bordes, *The Old Stone Age*, trans. J. E. Anderson (New York: McGraw-Hill Book Co., 1968), pp. 147–166; Henri Laville, Jean-Philippe Rigaud, and James Sackett, *Rock Shelters of the Perigord* (New York: Academic Press, 1980), pp. 218–227.

77. Hoffecker, *Desolate Landscapes*, pp. 158–162.

78. J. B. Birdsell, "The Recalibration of a Paradigm for the First Peopling of Greater Australia," in *Sunda and Sahul: Prehistoric Studies in Southeast Asia, Melanesia, and Australia*, ed. J. Allen, J. Golson, and R. Jones, pp. 113–167 (London: Academic Press, 1977).

79. Goebel, "The Pleistocene Colonization of Siberia," pp. 216–218; Hoffecker, *Desolate Landscapes*, p. 172. The oldest-known eyed needle was recovered from Kostënki 15 in Russia and is probably about 35,000 years old. See A. N. Rogachev and A. A. Sinitsyn, "Kostenki 15 (Gorodtsovskaya Stoyanka)," in *Paleolit Kostenkovsko-Borshchevskogo Raiona na Donu 1879–1979*, ed. N. D. Praslov and A. N. Rogachev, pp. 162–171 (Leningrad: Nauka, 1982). Fragments of needles (without eyes) of comparable age are reported from the Siberian site of Tolbaga. See Troeng, *Worldwide Chronology*, pp. 125–130. New evidence for the chronology of modern human clothing recently appeared from an unexpected source. An analysis of DNA sequences from a global sample of body lice *(Pediculus humanus)* indicated an African origin at roughly 72,000 years ago ($\pm 42,000$ years). Because the body louse inhabits clothing, its origin and age seem to be significant. Ralf Kittler, Manfred Kayser, and Mark Stoneking, "Molecular Evolution of *Pediculus humanus* and the Origin of Clothing." *Current Biology* 13 (2003): 1414–1417.

80. G. Medvedev, "Upper Paleolithic Sites in South-Central Siberia," in *The Palaeolithic of Siberia: New Discoveries and Interpretations*, ed. A. P. Derevianko, pp. 122–132 (Urbana: University of Illinois Press, 1998), p. 225, fig. 114.

81. S. A. Semenov, *Prehistoric Technology*, trans. M. W. Thompson (New York: Barnes and Noble, 1964), pp. 85–93; H. Juel Jensen, "Functional Analysis of Prehistoric Flint Tools by High-Powered Microscopy: A Review of West European Research," *Journal of World Prehistory* 2, no. 1 (1988): 53–88; Patricia Anderson-Gerfaud, "Aspects of Behaviour in the Middle Palaeolithic: Functional Analysis of Stone Tools from Southwest France," in *The Emergence of Modern Humans*, ed. P. Mellars, pp. 389–418 (Edinburgh: Edinburgh University Press, 1990), p. 405.

82. Goebel, "The Pleistocene Colonization of Siberia," pp. 213–218; Hoffecker, *Desolate Landscapes*, pp. 162–163.

83. Olga Soffer, J. M. Adovasio, and D. C. Hyland, "The 'Venus' Figurines: Textiles, Basketry, Gender, and Status in the Upper Paleolithic," *Current Anthropology* 41, no. 4 (2000): 511–537; Hoffecker, *Desolate Landscapes*, p. 161.

84. Michael P. Richards et al., "Stable Isotope Evidence for Increasing Dietary Breadth in the European Mid-Upper Paleolithic," *Proceedings of the National Academy of Sciences* 98 (2001): 6528–6532.

85. I. G. Shovkoplyas, *Mezinskaya Stoyanka* (Kiev: Naukova Dumka, 1965), pp. 203–206.

86. V. Gordon Childe, *Man Makes Himself* (London: Watts and Co., 1936), pp. 63–64.

87. Examples of mechanical food-getting technology in these and other categories among nonindustrial peoples are described in Oswalt, *An Anthropological Analysis of Food-Getting Technology*. The complex clothing of recent northern peoples is described in a classic 1914 monograph by Gudmund Hatt, "Arctic Skin Clothing in Eurasia and America: An Ethnographic Study," *Arctic Anthropology* 5, no. 2 (1969): 3–132. Mechanical fire-making equipment is illustrated in Carleton S. Coon, *The Hunting Peoples* (New York: Little, Brown and Co., 1971).

88. Pamela P. Vandiver et al., "The Origins of Ceramic Technology at Dolni Vestonice, Czechoslovakia," *Science* 246 (1989): 1002–1008; Olga Soffer et al., "The Pyrotechnology of Performance Art: Moravian Venuses and Wolverines," in *Before Lascaux: The Complex Record of the Early Upper Paleolithic*, ed. H. Knecht, A. Pike-Tay, and R. White, pp. 259–275 (Boca Raton, FL: CRC Press, 1993).

89. Velichko et al., "Periglacial Landscapes of the East European Plain"; A. A. Velichko, "Late Pleistocene Spatial Paleoclimatic Reconstructions," in *Late Quaternary Environments of the Soviet Union*, ed. Velichko, pp. 261–285.

90. Lawrence Guy Straus, "The Upper Paleolithic of Europe: An Overview," *Evolutionary Anthropology* 4, no. 1 (1995): 4–16; Goebel, "The Pleistocene Colonization of Siberia," pp. 216–218.

91. R. Dale Guthrie, *Frozen Fauna of the Mammoth Steppe: The Story of Blue Babe* (Chicago: University of Chicago Press, 1990); R. Dale Guthrie, "Origin and Causes of the Mammoth Steppe: A Story of Cloud Cover, Woolly Mammoth Tooth Pits, Buckles, and Inside-Out Beringia," *Quaternary Science Reviews* 20, nos. 1–3 (2001): 549–574; Hoffecker, *Desolate Landscapes*, pp. 22–26.

92. Parallels between the European Upper Paleolithic and modern peoples of the Arctic were observed in 1865 by Sir John Lubbock in the first edition of *Pre-historic Times*, and later by many others, including William J. Sollas, Miles Burkitt, V. Gordon Childe, and Grahame Clark. See Trigger, *A History of Archaeological Thought*, pp. 114–155; John F. Hoffecker, "The Eastern Gravettian 'Kostenki Culture' as an Arctic Adaptation," *Anthropological Papers of the University of Alaska*, n.s. 2, no. 1 (2002): 115–136.

93. Semenov, *Prehistoric Technology*, pp. 93–94; L. M. Tarasov, *Gagarinskaya Stoyanka i Ee Mesto v Paleolite Evropy* (Leningrad: Nauka, 1979); G. P. Grigor'ev, "The Kostenki-Avdeevo Archaeological Culture and the Willendorf-Pavlov-Kostenki-Avdeevo Cultural Unity," in *From Kostenki to Clovis*, ed. O. Soffer and N. D. Praslov, pp. 51–65 (New York: Plenum Press, 1993); Hoffecker, "The Eastern Gravettian 'Kostenki Culture.'"

94. Richard G. Klein, *Ice-Age Hunters of the Ukraine* (Chicago: University of Chicago Press, 1973), pp. 100–101; Hoffecker, *Desolate Landscapes*, pp. 226–227.

95. Richards et al., "Stable Isotope Evidence for Increasing Dietary Breadth"; Hoffecker, "The Eastern Gravettian 'Kostenki Culture.'"

96. Lewis R. Binford, "Willow Smoke and Dogs' Tails: Hunter-Gatherer Settlement Systems and Archaeological Site Formation," *American Antiquity* 45, no. 1 (1980): 4–20; Lewis R. Binford, "Mobility, Housing, and Environment: A Comparative Study," *Journal of Anthropological Research* 46 (1990): 119–152; Robin Torrence, "Time Budgeting and Hunter-Gatherer Technology," in *Hunter-Gatherer Economy in Prehistory: A European Perspective*, ed. G. Bailey, pp. 11–22 (Cambridge: Cambridge University Press, 1983).

97. Klein, *Ice-Age Hunters of the Ukraine*, pp. 100–104; Grigor'ev, "The Kostenki-Avdeevo Archaeological Culture"; Hoffecker, *Desolate Landscapes*, pp. 244–246. Similar feature complexes ("longhouses") are found in the North American Arctic. See Robert McGhee, *Ancient People of the Arctic* (Vancouver: University of British Columbia, 1996).

98. J. K. Kozlowski, "The Gravettian in Central and Eastern Europe," in *Advances in World Archaeology*, vol. 5, ed. F. Wendorf and A. E. Close, pp. 131–200 (Orlando, FL: Academic Press, 1986); Hoffecker, "The Eastern Gravettian 'Kostenki Culture.'"

99. V. P. Grichuk, "Late Pleistocene Vegetation History," in *Late Quaternary Environments of the Soviet Union*, ed. Velichko, pp. 155–178; Velichko, "Late Pleistocene Spatial Paleoclimatic Reconstructions," pp. 273–279. Unlike sites of Last Glacial Maximum age on the central East European Plain, most Siberian sites from this interval contain wood charcoal. See N. F. Lisitsyn and Yu. S. Svezhentsev, "Radiouglerodnaya Khronologiya Verkhnego Paleolita Severnoi Azii," in *Radiouglerodnaya Khronologiya Paleolita Vostochnoi Azii Problemy i Perspektivy*, ed. A. A. Sinitsyn and N. D. Praslov, pp. 67–108 (Saint Petersburg: Russian Academy of Sciences, 1997).

100. Chester S. Chard, *Northeast Asia in Prehistory* (Madison: University of Wisconsin Press, 1974), pp. 20–27; Medvedev, "Upper Paleolithic Sites in South-Central Siberia"; Goebel, "The Pleistocene Colonization of Siberia," pp. 216–218.

101. Robin Dennell, *European Economic Prehistory: A New Approach* (London: Academic Press, 1983), pp. 100–102; Olga Soffer, *The Upper Paleolithic of the Central Russian Plain* (San Diego: Academic Press, 1985), pp. 173–176.

102. Goebel, "The Pleistocene Colonization of Siberia," pp. 210–218; Dolukhanov, Sokoloff, and Shukurov, "Radiocarbon Chronology of Upper Palaeolithic Sites," p. 709, fig. 6; Hoffecker, *Desolate Landscapes*, pp. 200–201, fig. 6.3.

103. Straus, "The Upper Paleolithic of Europe," pp. 9–11.

104. Guthrie, *Frozen Fauna of the Mammoth Steppe*, pp. 239–245.

105. Trinkaus, "Neanderthal Limb Proportions"; Holliday, "Brachial and Crural Indices."

106. Hoffecker and Elias, "Environment and Archaeology in Beringia," pp. 37–38.

107. K. D. Orr and D. C. Fainer, "Cold Injuries in Korea during Winter of 1950–51," *Military Medicine* 31 (1952): 177–220; D. Miller and D. R. Bjornson, "An Investigation of Cold Injured Soldiers in Alaska," *Military Medicine* 127 (1962):

247–252; D. S. Sumner, T. L. Criblez, and W. H. Doolittle, "Host Factors in Human Frostbite," *Military Medicine* 139 (1974): 454–461.

108. The Siberian Yakuts lived in the Yana River area in Northeast Asia, where the mean January temperature is approximately −50° C. See Paul Lydolph, *Geography of the U.S.S.R.*, 3rd ed. (New York: John Wiley, 1977), p. 442.

Chapter 6. Into the Arctic

1. Almost all prehistory texts recognize a fundamental division between the cultures of the Pliocene and Pleistocene (that is, Paleolithic) and those of the Holocene (Mesolithic, Neolithic, Bronze Age, and so on). See, for example, V. Gordon Childe, *What Happened in History*, rev. ed. (Harmondsworth: Penguin Books, 1954); Chester S. Chard, *Northeast Asia in Prehistory* (Madison: University of Wisconsin Press, 1974); Grahame Clark, *World Prehistory in New Perspective*, 3rd ed. (Cambridge: University of Cambridge Press, 1977).

2. As discussed in chapter 5, the transition from the Middle to the Upper Paleolithic (that is, the dispersal of modern humans out of Africa) represents a more fundamental division in the archaeological record.

3. Lawrence Guy Straus, "The Upper Paleolithic of Europe: An Overview," *Evolutionary Anthropology* 4, no. 1 (1995): 4–16; Ted Goebel, "The Pleistocene Colonization of Siberia and Peopling of the Americas: An Ecological Approach," *Evolutionary Anthropology* 8 (1999): 208–227; John F. Hoffecker, *Desolate Landscapes: Ice-Age Settlement of Eastern Europe* (New Brunswick, NJ: Rutgers University Press, 2002), pp. 200–201.

4. Clive Gamble, *Timewalkers: The Prehistory of Global Colonization* (Cambridge: Harvard University Press, 1994), pp. 211–214; Vladimir Pitul'ko, "Terminal Pleistocene—Early Holocene Occupation in Northeast Asia and the Zhokhov Assemblage," *Quaternary Science Reviews* 20, nos. 1–3 (2001): 267–275.

5. T. W. Holliday, "Brachial and Crural Indices of European Late Upper Paleolithic and Mesolithic Humans," *Journal of Human Evolution* 36 (1999): 549–566; John F. Hoffecker, "The Eastern Gravettian 'Kostenki Culture' as an Arctic Adaptation," *Anthropological Papers of the University of Alaska*, n.s., 2, no. 1 (2002): 115–136.

6. John F. Hoffecker and Scott A. Elias, "Environment and Archeology in Beringia," *Evolutionary Anthropology* 12, no. 1 (2003): 34–49.

7. Moreau S. Maxwell, *Prehistory of the Eastern Arctic* (Orlando, FL: Academic Press, 1985), pp. 45–51.

8. Henri Laville, Jean-Philippe Rigaud, and James Sackett, *Rock Shelters of the Perigord* (New York: Academic Press, 1980), pp. 299–311.

9. Anta Montet-White, *Le Malpas Rockshelter* (University of Kansas Publications in Anthropology no. 4, 1973), pp. 41–58.

10. Pierre Cattelain, "Un Crochet de Propulseur Solutréen de la Grotte de Combe-Saunière I (Dordogne)," *Bulletin de la Société Préhistorique Française* 86 (1989): 213–216; Pierre Cattelain, "Hunting during the Upper Paleolithic: Bow, Spearthrower, or Both?" in *Projectile Technology*, ed. H. Knecht, pp. 213–240 (New

York: Plenum Press, 1997), pp. 214–215; Michael Jochim, "The Upper Palaeolithic," in *European Prehistory: A Survey*, ed. S. Milisauskas, pp. 55–113 (New York: Kluwer Academic/Plenum Publishers, 2002), pp. 99–100.

11. Clive Gamble, *The Palaeolithic Settlement of Europe* (Cambridge: University of Cambridge Press, 1986), p. 122; Cattelain, "Hunting during the Upper Paleolithic," pp. 220–221; Richard G. Klein, *The Human Career*, 2nd ed. (Chicago: University of Chicago Press, 1999), pp. 540–542.

12. Robin Dennell, *European Economic Prehistory: A New Approach* (London: Academic Press, 1983), p. 90; Straus, "The Upper Paleolithic of Europe," p. 10.

13. François Bordes, *The Old Stone Age*, trans. J. E. Anderson (New York: McGraw-Hill Book Co., 1968), pp. 161–166; Lawrence Guy Straus, *Iberia before the Iberians: The Stone Age Prehistory of Cantabrian Spain* (Albuquerque: University of New Mexico Press, 1992), pp. 140–145.

14. Straus, *Iberia before the Iberians*, pp. 90–166; Jochim, "The Upper Palaeolithic," pp. 84–98.

15. Straus, *Iberia before the Iberians*, pp. 159–193; Paul G. Bahn and Jean Vertut, *Journey through the Ice Age* (Berkeley: University of California Press, 1997); Jochim, "The Upper Palaeolithic," pp. 99–105.

16. Straus, "The Upper Paleolithic of Europe," pp. 9–11; Holliday, "Brachial and Crural Indices," p. 562.

17. Dennell, *European Economic Prehistory*, pp. 129–151.

18. Signe E. Nygaard, "The Stone Age of Northern Scandinavia: A Review," *Journal of World Prehistory* 3, no. 1 (1989): 71–116.

19. Ericka Helskog, "The Komsa Culture: Past and Present," *Arctic Anthropology* 11, suppl. (1974): 261–265; Ericka Engelstad, "Mesolithic House Sites in Arctic Norway," in *The Mesolithic in Europe*, ed. C. Bonsall, pp. 331–337 (Edinburgh: John Donald Publishers, 1989); Nygaard, "The Stone Age of Northern Scandinavia," pp. 81–82; N. N. Gurina, "Mezolit Kol'skogo Poluostrova," in *Mezolit SSSR*, ed. L. V. Kol'tsov, pp. 20–26 (Moscow: Nauka, 1989).

20. Ørnulv Vorren, *Norway North of 65* (Oslo: Oslo University Press, 1960); Ericka Engelstad, "The Late Stone Age of Arctic Norway: A Review," *Arctic Anthropology* 22, no. 1 (1985): 79–96. In terms of climate, the "subarctic" has been defined as a region where mean temperature is 50° F (10° C) or higher for at least one but no more than four months during the year. See Steven B. Young, *To the Arctic: An Introduction to the Far Northern World* (New York: John Wiley and Sons, 1994), pp. 15–17.

21. Gutorm Gjessing, "Circumpolar Stone Age," *Acta Arctica* 2 (1944): 1–70; Gutorm Gjessing, "The Circumpolar Stone Age," *Antiquity* 27 (1953): 131–136; Erik Trinkaus, "Neanderthal Limb Proportions and Cold Adaptation," in *Aspects of Human Evolution*, ed. C. Stringer, pp. 187–224 (London: Taylor and Francis, 1981); G. Richard Scott et al., "Physical Anthropology of the Arctic," in *The Arctic: Environment, People, Policy*, ed. M. Nuttall and T. V. Callaghan, pp. 339–373 (Amsterdam: Harwood Academic Publishers, 2000).

22. Gutorm Gjessing, "Maritime Adaptations in Northern Norway's Prehistory," in *Prehistoric Maritime Adaptations of the Circumpolar Zone*, ed. W. Fitzhugh, pp. 87–

100 (The Hague: Mouton Publishers, 1975); Engelstad, "Mesolithic House Sites in Arctic Norway."

23. In many areas, the warm oscillation that took place approximately 20,000 years ago is represented by a thin and weakly developed soil horizon that underlies the post–cold maximum sites of the central East European Plain. See Olga Soffer, *The Upper Paleolithic of the Central Russian Plain* (San Diego: Academic Press, 1985), pp. 232–233; Hoffecker, *Desolate Landscapes,* pp. 200–212.

24. A. A. Velichko et al., "Periglacial Landscapes of the East European Plain," in *Late Quaternary Environments of the Soviet Union,* ed. A. A. Velichko, pp. 94–118 (Minneapolis: University of Minnesota Press, 1984), pp. 110–117; A. A. Velichko, "Loess-Paleosol Formation on the Russian Plain," *Quaternary International* 7/8 (1990): 103–114.

25. J. K. Kozlowski, "The Gravettian in Central and Eastern Europe," in *Advances in World Archaeology,* vol. 5, ed. F. Wendorf and A. E. Close, pp. 131–200 (Orlando, FL: Academic Press, 1986); Jiří Svoboda, Vojen Ložek, and Emanuel Vlček, *Hunters between East and West: The Paleolithic of Moravia* (New York: Plenum Press, 1996), p. 143.

26. Holliday, "Brachial and Crural Indices"; Hoffecker, *Desolate Landscapes,* pp. 215–218.

27. Although isolated examples of mammoth-bone structures may date to before the cold maximum, most—if not all—of them postdate 20,000 years ago. See I. G. Pidoplichko, *Pozdnepaleoliticheskie Zhilishcha iz Kostei Mamonta na Ukraine* (Kiev: Naukova Dumka, 1969); Soffer, *The Upper Paleolithic of the Central Russian Plain;* Hoffecker, *Desolate Landscapes,* pp. 231–232.

28. For example, see J. Louis Giddings, *Ancient Men of the Arctic* (New York: Alfred A. Knopf, 1967), pp. 73–97.

29. Pidoplichko, *Pozdnepaleoliticheskie Zhilishcha iz Kostei Mamonta na Ukraine;* Richard G. Klein, *Ice-Age Hunters of the Ukraine* (Chicago: University of Chicago Press, 1973), pp. 91–99; M. I. Gladkih, N. L. Kornietz, and O. Soffer, "Mammoth-Bone Dwellings on the Russian Plain," *Scientific American* 251, no. 5 (1984): 164–175; Z. A. Abramova, "Two Examples of Terminal Paleolithic Adaptations," in *From Kostenki to Clovis,* ed. O. Soffer and N. D. Praslov, pp. 85–100 (New York: Plenum Press, 1993), pp. 86–95.

30. Soffer, *The Upper Paleolithic of the Central Russian Plain.*

31. Hoffecker, *Desolate Landscapes,* pp. 246–247.

32. Klein, *Ice-Age Hunters of the Ukraine,* pp. 88–89; Soffer, *The Upper Paleolithic of the Central Russian Plain,* pp. 371–372.

33. Water-sieving of sediment samples at Eliseevichi yielded fish remains in the form of eye lenses (N. K. Vereshchagin and I. E. Kuz'mina, "Ostatki Mlekopitayushchikh iz Paleoliticheskikh Stoyanok na Donu i Verkhnei Desne," *Trudy Zoologicheskogo Instituta AN SSSR* 72 (1977): 77–110), while isolated bones of pike were recovered from Kostenki and Mezhirich (P. I. Boriskovskii, *Ocherki po Paleolitu Basseina Dona,* Materialy i Issledovaniya po Arkheologii SSSR 121 (1963), p. 78; N. L. Korniets et al., "Mezhirich," in *Arkheologiya i Paleogeografiya Pozdnego Paleolita Russkoi Ravniny,* ed. I. P. Gerasimov, pp. 106–119 (Moscow: Nauka, 1981), p. 115). Bird

remains (especially willow ptarmigan) have been recovered from many sites. See Klein, *Ice-Age Hunters of the Ukraine,* p. 57.

34. I. G. Pidoplichko, *Mezhirichskie Zhilishcha iz Kostei Mamonta* (Kiev: Naukova Dumka, 1976), p. 165; Hoffecker, *Desolate Landscapes,* pp. 228–232.

35. Mikhail V. Sablin and Gennady A. Khlopachev, "The Earliest Ice Age Dogs: Evidence from Eliseevichi I," *Current Anthropology* 43, no. 5 (2002): 795–799; Christy G. Turner, "Teeth, Needles, Dogs, and Siberia: Bioarchaeological Evidence for the Colonization of the New World," in *The First Americans,* pp. 123–158 (Memoirs of the California Academy of Sciences no. 27, 2002), pp. 145–146.

36. Alexander Marshack, "Upper Paleolithic Symbol Systems of the Russian Plain: Cognitive and Comparative Analysis," *Current Anthropology* 20 (1979): 271–311; Kozlowski, "The Gravettian in Central and Eastern Europe," pp. 183–184.

37. L. V. Kol'tsov, *Final'nyi Paleolit i Mezolit Yuzhnoi i Vostochnoi Pribaltiki* (Moscow: Nauka, 1977), pp. 120–135; Janusz Kozlowski and H.-G. Bandi, "The Paleohistory of Circumpolar Arctic Colonization," *Arctic* 37, no. 4 (1984): 359–372.

38. P. M. Dolukhanov and N. A. Khotinskiy, "Human Cultures and Natural Environment in the USSR during the Mesolithic and Neolithic," in *Late Quaternary Environments of the Soviet Union,* ed. Velichko, pp. 319–327; N. A. Khotinskiy, "Holocene Climatic Change," in *Late Quaternary Environments of the Soviet Union,* ed. Velichko, pp. 305–309.

39. Grigoriy M. Burov, "Some Mesolithic Wooden Artifacts from the Site of Vis I in the European North East of the U.S.S.R.," in *The Mesolithic in Europe,* ed. C. Bonsall, pp. 391–401 (Edinburgh: John Donald Publishers, 1989); S. V. Oshibkina, "The Material Culture of the Veretye-type Sites in the Region to the East of Lake Onega," in *The Mesolithic in Europe,* ed. Bonsall, pp. 402–413; Ilga Zagorska and Francis Zagorskis, "The Bone and Antler Inventory from Zvejnieki II, Latvian SSR," in *The Mesolithic in Europe,* ed. Bonsall, pp. 414–423. In a classic monograph, Cornelius Osgood, *Ingalik Material Culture* (Yale University Publications in Anthropology no. 22, 1940), described in substantial detail the material culture of the Ingalik—an Athapaskan group living in the lower Yukon River area who obtained 50 percent of their diet from fishing and the remainder from hunting (40 percent) and some plant gathering (10 percent). See Robert L. Kelly, *The Foraging Spectrum: Diversity in Hunter-Gatherer Lifeways* (Washington, DC: Smithsonian Institution Press, 1995), p. 67, table 3-1.

40. Grahame Clark and Stuart Piggot, *Prehistoric Societies* (Harmondsworth: Penguin Books, 1970), p. 136.

41. Burov, "Some Mesolithic Wooden Artifacts," pp. 393–400; Oshibkina, "The Material Culture of the Veretye-type Sites," pp. 408–410.

42. Burov, "Some Mesolithic Wooden Artifacts," pp. 393–397.

43. J.G.D. Clark, "Neolithic Bows from Somerset, England, and the Prehistory of Archery in North-west Europe," *Proceedings of the Prehistoric Society* 29 (1963): 50–98; Oshibkina, "The Material Culture of the Veretye-type Sites," p. 410.

44. S. V. Oshibkina, "Mezolit Tsentral'nykh i Severo-Vostochnykh Raionov Severa Evropeiskoi Chasti SSSR," in *Mezolit SSSR,* ed. L. V. Kol'tsov, pp. 32–45 (Moscow: Nauka, 1989), pp. 34–37.

45. N. N. Gurina, "Mezolit Karelii," in *Mezolit SSSR*, ed. L. V. Kol'tsov, pp. 27–31; Oshibkina, "The Material Culture of the Veretye-type Sites," pp. 411–412.

46. Kenneth H. Jacobs, "Climate and the Hominid Postcranial Skeleton in Wurm and Early Holocene Europe," *Current Anthropology* 26 (1985): 512–514; Holliday, "Brachial and Crural Indices," pp. 562–563.

47. Oshibkina, "Mezolit Tsentral'nykh i Severo-Vostochnykh Raionov Severa Evropeiskoi Chasti SSSR," pp. 44–45.

48. Goebel, "The Pleistocene Colonization of Siberia," pp. 218–220; Ted Goebel et al., "Studenoe-2 and the Origins of Microblade Technologies in the Transbaikal, Siberia," *Antiquity* 74 (2000): 567–575.

49. Ted Goebel, "The 'Microblade Adaptation' and Recolonization of Siberia during the Late Upper Pleistocene," in *Thinking Small: Global Perspectives on Microlithization*, ed. R. G. Elston and S. L. Kuhn, pp. 117–131 (Archaeological Papers of the American Anthropological Association no. 12, 2002), pp.123–124.

50. V. P. Grichuk, "Late Pleistocene Vegetation History," in *Late Quaternary Environments of the Soviet Union*, ed. Velichko, pp. 155–178; Goebel, "The 'Microblade Adaptation,'" p. 125, fig. 9.4. Red deer *(Cervus elaphus)* is extremely rare in central East European Plain sites of this interval and elk or moose *(Alces alces)* is completely absent (Hoffecker, *Desolate Landscapes*, pp. 239–242).

51. Goebel, "The 'Microblade Adaptation,'" pp. 123–126.

52. Chard, *Northeast Asia in Prehistory*, p. 32, fig. 1.18; Z. A. Abramova, *Paleolit Eniseya: Kokorevskaya Kul'tura* (Novosibirsk: Nauka, 1979); V. F. Gening and V. T. Petrin, *Pozdnepaleolitcheskaya Epokha na Yuge Zapadnoi Sibiri* (Novosibirsk: Nauka, 1985), p. 48, fig. 17.

53. Goebel, "The 'Microblade Adaptation,'" p. 124.

54. Abramova, *Paleolit Eniseya: Kokorevskaya Kul'tura;* Z. A. Abramova, *Paleolit Eniseya: Afontovskaya Kul'tura* (Novosibirsk: Nauka, 1979); Sergey A. Vasil'ev, "The Final Paleolithic in Northern Asia: Lithic Assemblage Diversity and Explanatory Models," *Arctic Anthropology* 38, no. 2 (2001): 3–30.

55. N. M. Ermolova, *Teriofauna Doliny Angary v Pozdnem Antropogene* (Novosibirsk: Nauka, 1978), pp. 26–33; Goebel, "The 'Microblade Adaptation,'" pp. 124–126.

56. M. P. Gryaznov, "Ostatki Cheloveka iz Kul'turnogo Sloya Afontova Gory," *Trudy Komissii po Izucheniyu Chetvertichnogo Perioda* 1 (1932): 137–144; V. P. Alekseev, "The Physical Specificities of Paleolithic Hominids in Siberia," in *The Paleolithic of Siberia: New Discoveries and Interpretations*, ed. A. P. Derevianko, pp. 329–335 (Urbana: University of Illinois Press, 1998), pp. 329–330; Hoffecker and Elias, "Environment and Archeology in Beringia," p. 37–38.

57. M. P. Aksenov, "Archaeological Investigations at the Stratified Site of Verkholenskaia Gora in 1963–1965," *Arctic Anthropology* 6, no. 1 (1969): 74–87; Ermolova, *Teriofauna Doliny Angary v Pozdnem Antropogene*, pp. 31–33; Goebel, "The 'Microblade Adaptation,'" p. 126.

58. G. I. Medvedev, "Results of the Investigations of the Mesolithic in the Stratified Settlement of Ust-Belaia 1957–1964," *Arctic Anthropology* 6, no. 1 (1969): 61–73; Ermolova, *Teriofauna Doliny Angary v Pozdnem Antropogene*, pp. 34–40.

59. Yu. A. Mochanov, *Drevneishie Etapy Zaseleniya Chelovekom Severo-Vostochoi Azii* (Novosibirsk: Nauka, 1977); pp. 6–31; Yuri A. Mochanov and Svetlana A. Fedoseeva, "Dyuktai Cave," in *American Beginnings: The Prehistory and Palaeoecology of Beringia*, ed. F. H. West, pp. 164–174 (Chicago: University of Chicago Press, 1996).

60. Mochanov, *Drevneishie Etapy Zaseleniya*, pp. 241–253; Vladimir Pitul'ko, "Terminal Pleistocene—Early Holocene Occupation in Northeast Asia and the Zhokhov Assemblage," *Quaternary Science Reviews* 20, nos. 1–3 (2001): 267–275.

61. William Roger Powers, "Paleolithic Man in Northeast Asia," *Arctic Anthropology* 10, no. 2 (1973): 1–106; Mochanov, *Drevneishie Etapy Zaseleniya*, pp. 223–240.

62. See O. W. Archibold, *Ecology of World Vegetation* (London: Chapman and Hall, 1995), p. 8, fig. 1.6.

63. Mochanov, *Drevneishie Etapy Zaseleniya*, pp. 248–249; L. V. Kol'tsov, "Mezolit Severa Sibiri i Dal'nego Vostoka," in *Mezolit SSSR*, ed. L. V. Kol'tsov, pp. 187–194 (Moscow: Nauka, 1989), pp. 187–191.

64. In terms of climate (as opposed to latitude), the Arctic has been defined as a region where the mean temperature remains below 50° F (10° C) for all months of the year. See Young, *To the Arctic*, pp. 13–18.

65. Vladimir V. Pitul'ko, "An Early Holocene Site in the Siberian High Arctic," *Arctic Anthropology* 30, no. 1 (1993): 13–21; Pitul'ko, "Terminal Pleistocene—Early Holocene Occupation," p. 270; Ted Goebel and Sergei B. Slobodin, "The Colonization of Western Beringia: Technology, Ecology, and Adaptation," in *Ice Age Peoples of North America: Environments, Origins, and Adaptations of the First Americans*, ed. R. Bonnichsen and K. L. Turnmire, pp. 104–155 (Corvallis: Oregon State University Press, 1999).

66. Pitul'ko, "An Early Holocene Site," pp. 19–20; V. V. Pitul'ko and A. K. Kasparov, "Ancient Arctic Hunters: Material Culture and Survival Strategy," *Arctic Anthropology* 33 (1996): 1–36.

67. L. P. Khlobystin, "O Drevnem Zaselenii Arktiki," *Kratkie Soobshcheniya Instituta Arkheologii* 36 (1973): 11–16; L. P. Khlobystin and G. M. Levkovskaya, "Rol' Sotsial'nogo i Ekologicheskogo Faktorov v Razvitii Arkticheskikh Kul'tur Evrazii," in *Pervobytnyi Chelovek, Ego Material'naya Kul'tura i Prirodnaya Sreda v Pleistotsene i Golotsene*, ed. I. P. Gerasimov, pp. 235–242 (Moscow: USSR Academy of Sciences, 1974), pp. 238–239; Kol'tsov, "Mezolit Severa Sibiri i Dal'nego Vostoka," p. 192.

68. Troy L. Péwé, *Quaternary Geology of Alaska* (Geological Survey Professional Paper 835, 1975), pp. 95–101; R. Dale Guthrie, *Frozen Fauna of the Mammoth Steppe: The Story of Blue Babe* (Chicago: University of Chicago, Press, 1990), pp. 45–80.

69. R. Dale Guthrie, "Paleoecology of the Large-Mammal Community in Interior Alaska during the Late Pleistocene," *American Midland Naturalist* 79, no. 2 (1968): 346–363; Guthrie, *Frozen Fauna of the Mammoth Steppe*, pp. 239–245; J. V. Matthews, "East Beringia during Late Wisconsin Time: A Review of the Biotic Evidence," in *Paleoecology of Beringia*, ed. D. M. Hopkins et al., pp. 127–150 (New York: Academic Press, 1982), pp. 139–143.

70. Eric Hultén, *Outline of the History of Arctic and Boreal Biota during the Quaternary Period* (Stockholm: Bokförlags Aktiebolaget Thule, 1937); Guthrie, *Frozen Fauna of the Mammoth Steppe*, pp. 205–225; R. Dale Guthrie, "Origin and Causes of

the Mammoth Steppe: A Story of Cloud Cover, Woolly Mammoth Tooth Pits, Buckles, and Inside-Out Beringia," *Quaternary Science Reviews* 20, nos. 1–3 (2001): 549–574.

71. Guthrie, *Frozen Fauna of the Mammoth Steppe* pp. 273–277; Hoffecker and Elias, "Environment and Archeology in Beringia," pp. 36–41.

72. Richard E. Morlan and Jacques Cinq-Mars, "Ancient Beringians: Human Occupation in the Late Pleistocene of Alaska and the Yukon Territory," in *Paleoecology of Beringia*, ed. Hopkins et al., pp. 353–381. Charles E. Holmes, "Broken Mammoth," in *American Beginnings*, ed. West, pp. 312–318; Charles E. Holmes, Richard VanderHoek, and Thomas E. Dilley, "Swan Point," in *American Beginnings*, ed. West, pp. 319–323; David R. Yesner, "Human Dispersal into Interior Alaska: Antecedent Conditions, Mode of Colonization, and Adaptations," *Quaternary Science Reviews* 20, nos. 1–3 (2001): 315–327. The site of Berelëkh, located near the mouth of the Indigirka River at latitude 70° North, is often listed as one of the earliest occupations east of the Lena Basin (and one of the oldest sites at this latitude), but uncertainties surround the dates—reported at ca. 16,000–15,500 years ago but possibly much younger. See Mochanov, *Drevneishie Etapy Zaseleniya*, pp. 76–87; Pitul'ko, "Terminal Pleistocene–Early Holocene Occupation," p. 267; Hoffecker and Elias, "Environment and Archeology in Beringia," pp. 39–44. A recently reported site near the Yana River Mouth (mentioned in chapter 5) dates to almost 30,000 years ago and suggests an even earlier appearance east of the Lena Basin at this latitude—at least on a seasonal basis—when climates were somewhat warmer toward the end of the Middle Pleniglacial. See V. V. Pitulko et al., "The Yana RHS Site: Humans in the Arctic before the Last Glacial Maximum." *Science* 303 (2004): 52–56. Access to the New World was apparently blocked at this time by glacial ice sheets. See Carole A. S. Mandryk et al., "Late Quaternary Paleoenvironments of Northwestern North America: Implications for Inland versus Coastal Migration Routes." *Quaternary Science Reviews* 20, nos. 1–3 (2001): 301–314.

73. E. James Dixon, "Human Colonization of the Americas: Timing, Technology, and Process," *Quaternary Science Reviews* 20, nos. 1–3 (2001): 277–299; Mandryk et al., "Late Quaternary Paleoenvironments of Northwestern North America, 301–314; T. D. Dillehay and M. B. Collins, "Early Cultural Evidence from Monte Verde in Chile," *Nature* 332 (1988): 150–152.

74. Nancy H. Bigelow and Wm. Roger Powers, "Climate, Vegetation, and Archaeology 14,000–9000 cal yr B.P. in Central Alaska," *Arctic Anthropology* 38, no. 2 (2001): 171–195; Hoffecker and Elias, "Environment and Archeology in Beringia," pp. 40–44.

75. Nikolai N. Dikov, "The Ushki Sites, Kamchatka Peninsula," in *American Beginnings*, ed. West, pp. 244–250; Ted Goebel, Michael R. Waters, and Margarita Dikova, "The Archaeology of Ushki Lake, Kamchatka, and the Pleistocene Peopling of the Americas," *Science* 301 (2003): 501–505.

76. William R. Powers and John F. Hoffecker, "Late Pleistocene Settlement in the Nenana Valley, Central Alaska," *American Antiquity* 54, no. 2 (1989): 263–287; John F. Hoffecker, "Late Pleistocene and Early Holocene Sites in the Nenana River Valley, Central Alaska," *Arctic Anthropology* 38, no. 2 (2001): 139–153.

77. See, for example, John F. Hoffecker, W. Roger Powers, and Ted Goebel, "The Colonization of Beringia and the Peopling of the New World," *Science* 259 (1993): 46–53; Dixon, "Human Colonization of the Americas."

78. Frederick Hadleigh West, ed., *American Beginnings: The Prehistory and Palaeoecology of Beringia* (Chicago: University of Chicago Press, 1996); Owen K. Mason, Peter M. Bowers, and David M. Hopkins, "The Early Holocene Milankovitch Thermal Maximum and Humans: Adverse Conditions for the Denali Complex of Eastern Beringia," *Quaternary Science Reviews* 20, nos. 1–3 (2001): 525–548.

79. N. N. Dikov, *Arkheologicheskie Pamyatniki Kamchatki, Chukotki i Verkhnei Kolymy* (Moscow: Nauka, 1977), pp. 52–58; Dikov, "The Ushki Sites, Kamchatka Peninsula," pp. 245–246; Holmes, "Broken Mammoth," p. 317; Yesner, "Human Dispersal into Interior Alaska," pp. 321–322.

80. Scott A. Elias, Susan K. Short, and R. Lawrence Phillips, "Paleoecology of Late-Glacial Peats from the Bering Land Bridge, Chukchi Sea Shelf Region, Northwestern Alaska," *Quaternary Research* 38 (1992): 371–378; Bigelow and Powers, "Climate, Vegetation, and Archaeology 14,000–9000 cal yr B.P."; Scott A. Elias, "Beringian Paleoecology: Results from the 1997 Workshop," *Quaternary Science Reviews* 20, nos. 1–3 (2001): 7–13.

81. Frederick Hadleigh West, *The Archaeology of Beringia* (New York: Columbia University Press, 1981), pp. 91–154; Mason, Bowers, and Hopkins, "The Early Holocene Milankovitch Thermal Maximum and Humans," pp. 526–535.

82. Mason, Bowers, and Hopkins, "The Early Holocene Milankovitch Thermal Maximum and Humans," pp. 539–542. The steppe bison *(Bison priscus)* did not become extinct in parts of interior Alaska at the end of the Pleistocene and was hunted by Denali people who occupied sites in the Tanana Valley and northern foothills of the Alaska Range (Powers and Hoffecker, "Late Pleistocene Settlement in the Nenana Valley, Central Alaska," pp. 272–273; Yesner, "Human Dispersal into Interior Alaska," p. 321).

83. Severe coastal erosion would have destroyed sites of Denali age in some areas along the Beaufort Sea coast, but probably not on portions of the northwest coast.

84. Allen P. McCartney and Douglas W. Veltre, "Anangula Core and Blade Site," in *American Beginnings*, ed. West, pp. 443–450; Dixon, "Human Colonization of the Americas."

Chapter 7. Peoples of the Circumpolar Zone

1. R. Dale Guthrie, "Mammals of the Mammoth Steppe as Paleoenvironmental Indicators," in *Paleoecology of Beringia*, ed. D. M. Hopkins et al., pp. 307–326 (New York: Academic Press, 1982), pp. 315–320.

2. Robert H. Whittaker, *Communities and Ecosystems*, 2nd ed. (New York: Macmillan Publishing Co., 1975), pp. 212–213; Eugene P. Odum, *Ecology and Our Endangered Life-Support Systems* (Sunderland, MA: Sinauer Associates, 1993).

3. See, for example, Ericka Engelstad, "The Late Stone Age of Arctic Norway: A Review," *Arctic Anthropology* 22, no. 1 (1985): 79–96.

4. William Fitzhugh, "A Comparative Approach to Northern Maritime Adaptations," in *Prehistoric Maritime Adaptations of the Circumpolar Zone,* ed. W. Fitzhugh, pp. 339–386 (The Hague: Mouton Publishers, 1975); Don E. Dumond, *The Eskimos and Aleuts,* rev. ed. (London: Thames and Hudson, 1987).

5. William S. Laughlin, "Aleuts: Ecosystem, Holocene History, and Siberian Origin," *Science* 189 (1975): 507–515; Robert E. Ackerman, "Settlements and Sea Mammal Hunting in the Bering-Chukchi Sea Region," *Arctic Anthropology* 25, no. 1 (1988): 52–79; E. James Dixon, *Bones, Boats, and Bison* (Albuquerque: University of New Mexico Press, 1999), pp. 117–119.

6. Gutorm Gjessing, "Maritime Adaptations in Northern Norway's Prehistory," in *Prehistoric Maritime Adaptations of the Circumpolar Zone,* ed. Fitzhugh, pp. 87–100; Ericka Engelstad, "Mesolithic House Sites in Arctic Norway," in *The Mesolithic in Europe,* ed. C. Bonsall, pp. 331–337 (Edinburgh: John Donald Publishers, 1989), pp. 335–336.

7. Ericka Engelstad, "The Late Stone Age of Arctic Norway: A Review," *Arctic Anthropology* 22, no. 1 (1985): 79–96; Signe E. Nygaard, "The Stone Age of Northern Scandinavia: A Review," *Journal of World Prehistory* 3, no. 1 (1989): 71–116.

8. Anders Hagen, *Norway* (London: Thames and Hudson, 1967), pp. 68–77; Haakon Olsen, "Osteologisk Materiale, Innledning: Fisk-Fugl. Varangerfunnene VI," *Tromsø Museums Skrifter* 7, no. 6 (1967); Fitzhugh, "Comparative Approach," pp. 356–357.

9. Povl Simonsen, "When and Why Did Occupational Specialization Begin at the Scandinavian North Coast?" in *Prehistoric Maritime Adaptations of the Circumpolar Zone,* ed. Fitzhugh, pp. 75–85; M.A.P. Renouf, "Northern Coastal Hunter-Fishers: An Archaeological Model," *World Archaeology* 16, no. 1 (1984): 18–27.

10. Povl Simonsen, "Varanger-Funnene II. Fund og Udgravninger på Fjordens Sydkyst," *Tromsø Museums Skrifter* 7, no. 2 (1961); Hagen, *Norway,* pp. 73–75; Knut Helskog, "Boats and Meaning: A Study of Change and Continuity in the Alta Fjord, Arctic Norway, from 4200 to 500 Years B.C.," *Journal of Anthropological Archaeology* 4 (1985): 177–205; Gjessing, "Maritime Adaptations in Northern Norway's Prehistory," p. 90.

11. Helskog, "Boats and Meaning"; Nygaard, "The Stone Age of Northern Scandinavia," pp. 97–98.

12. Hagen, *Norway,* pp. 69–74.

13. N. N. Gurina, "Neolit Lesnoi i Lesostepnoi Zon Evropeiskoi Chasti SSSR," in *Kamennyi Vek na Territorii SSSR,* ed. A. A. Formozov, pp. 134–156 (Moscow: Nauka, 1970); Tadeusz Sulimirski, *Prehistoric Russia: An Outline* (London: John Baker, 1970); N. N. Gurina, "O Nekotopykh Obshchikh Elementakh Kul'tury Drevnikh Plemen Kol'skogo Poluostrova i ikh Sosedei," in *Paleolit i Neolit,* ed. V. P. Liubin, pp. 83–92 (Leningrad: Nauka, 1986).

14. Hagen, *Norway,* pp. 133–134; Engelstad, "The Late Stone Age of Arctic Norway," pp. 88–90; Sulimirski, *Prehistoric Russia,* pp. 328–329.

15. E. D. Prokof'yeva, "The Nentsy," in *The Peoples of Siberia,* ed. M. G. Levin and L. P. Potapov, pp. 547–570 (Chicago: University of Chicago Press, 1964); Roberto Bosi, *The Lapps* (London: Thames and Hudson, 1960); Aage Solbakk, ed., *The Sami*

People (Karasjok, Norway: Sámi Instituhtta/Davvi Girji O.S., 1990); Knut Odner, *The Varanger Saami: Habitation and Economy AD 1200–1900* (Oslo: Scandinavian University Press, 1992).

16. Bruce G. Trigger, *A History of Archaeological Thought* (Cambridge: Cambridge University Press, 1989), pp. 73–147.

17. As noted earlier, Late Stone Age people in arctic Norway seem to have largely abandoned pottery about 5,000 years ago, although they later readopted it (see Engelstad, "The Late Stone Age of Arctic Norway," pp. 82–83). Another example is the Ipiutak of northwest Alaska, who were undoubtedly descended from pottery-making people but had no use for ceramics (see Dumond, *The Eskimos and Aleuts*, pp. 114–118).

18. Yu. A. Mochanov, *Arkheologicheskie Pamyatniki Yakutii: Basseiny Aldana i Olekmy* (Novosibirsk: Nauka, 1983), pp. 16–17; Ackerman, "Settlements and Sea Mammal Hunting," p. 23.

19. Mochanov, *Arkheologicheskie Pamyatniki Yakutii*, p. 17; Yu. A. Mochanov et al., *Arkheologicheskie Pamyatniki Yakutii: Basseiny Vilyuya, Anabara i Oleneka* (Moscow: Nauka, 1991), p. 16.

20. Mochanov, *Arkheologicheskie Pamyatniki Yakutii*, p. 17.

21. Yu. A. Mochanov, "The Bel'kachinsk Neolithic Culture on the Aldan," *Arctic Anthropology* 6, no. 1 (1969): 104–114; Ackerman, "Settlements and Sea Mammal Hunting," p. 23.

22. Mochanov, "The Bel'kachinsk Neolithic Culture," pp. 111–113.

23. Chester S. Chard, *Northeast Asia in Prehistory* (Madison: University of Wisconsin Press, 1974), p. 74; W. R. Powers and R. H. Jordan, "Human Biogeography and Climate Change in Siberia and Arctic North America in the Fourth and Fifth Millennia BP," *Philosophical Transactions of the Royal Society London A* 330 (1990): 665–670.

24. J. Louis Giddings, *The Archeology of Cape Denbigh* (Providence, RI: Brown University Press, 1964), pp. 191–266; J. Louis Giddings, *Ancient Men of the Arctic* (New York: Alfred A. Knopf, 1967), pp. 246–276. A recent date of more than 5,000 years has been reported on a Denbigh Flint Complex occupation by Roger K. Harritt, "Paleo-Eskimo Beginnings in North America: A New Discovery at Kuzitrin Lake, Alaska," *Etudes/Inuit/Studies* 22, no. 1 (1998): 61–81.

25. Yu. A. Mochanov, "The Ymyiakhtakh Late Neolithic Culture," *Arctic Anthropology* 6, no. 1 (1969): 115–118. Some evidence of a primitive maritime economy dating to roughly 3,500 years ago has been reported from Wrangel Island and is discussed later. See N. N. Dikov, *Drevnie Kul'tury Severo-Vostochnoi Azii* (Moscow: Nauka, 1979), pp. 165–168. Much of the arctic coast of western and central Siberia remained without a maritime economy in historic times. The Nganasans, Yukaghir, and other native peoples of this region subsisted on terrestrial mammals, birds, and fish, and herded reindeer. See M. G. Levin and L. P. Potapov, eds., *The Peoples of Siberia*, trans. Stephen Dunn (Chicago: University of Chicago Press, 1964).

26. Robert McGhee, *Canadian Arctic Prehistory* (Hull, Quebec: Canadian Museum of Civilization, 1990), pp. 4–12.

27. Douglas D. Anderson, "A Stone Age Campsite at the Gateway to America," *Scientific American* 218, no. 6 (1968): 24–33; E. James Dixon, "Cultural Chronology of Central Interior Alaska," *Arctic Anthropology* 22, no. 1 (1985): 47–66; Dumond, *The Eskimos and Aleuts*, pp. 47–54.

28. Giddings, *The Archeology of Cape Denbigh*, p. 233; Dumond, *The Eskimos and Aleuts*, pp. 79–93.

29. Giddings, *The Archeology of Cape Denbigh*, pp. 229–233; Moreau S. Maxwell, *Prehistory of the Eastern Arctic* (Orlando, FL: Academic Press, 1985), pp. 77–98; Dumond, *The Eskimos and Aleuts*, pp. 80–84. Giddings suggested that the size range of the end-blades "suggests their use as sealing harpoon heads meant for the taking of both small seals and the big walrus or bearded seal."

30. Dennis J. Stanford, *The Walakpa Site, Alaska: Its Place in the Birnirk and Thule Cultures* (Smithsonian Contributions to Anthropology no. 20, 1976), pp. 10–17; Dumond, *The Eskimos and Aleuts*, pp. 82–85; Douglas D. Anderson, "Prehistory of North Alaska," in *Handbook of the North American Indian*, vol. 5, *Arctic*, ed. D. Damas, pp. 80–93 (Washington, DC: Smithsonian Institution, 1984); Douglas D. Anderson, *Onion Portage: The Archaeology of a Stratified Site from the Kobuk River, Northwest Alaska* (Anthropological Papers of the University of Alaska, vol. 22, nos. 1–2, 1988), pp. 73–102.

31. Hans-Georg Bandi, *Eskimo Prehistory*, trans. A. E. Keep (College: University of Alaska Press, 1969), pp. 157–161; Robert McGhee, *Ancient People of the Arctic* (Vancouver: University of British Columbia, 1996), pp. 30–72.

32. Anderson, "Prehistory of North Alaska," p. 84; McGhee, *Canadian Arctic Prehistory*, pp. 29–37.

33. Maxwell, *Prehistory of the Eastern Arctic*, pp. 61–62.

34. Eigil Knuth, "Archaeology of the Musk-Ox Way," *Contributions du Centre d'Etudes Arctiques et Finno-Scandinaves* 5 (1967); McGhee, *Ancient People of the Arctic*, p. 64.

35. McGhee, *Canadian Arctic Prehistory*, pp. 37–51.

36. Ibid., pp. 45–48.

37. Bandi, *Eskimo Prehistory*, pp. 136–139; Maxwell, *Prehistory of the Eastern Arctic*, pp. 84–95.

38. McGhee, *Canadian Arctic Prehistory*, p. 40.

39. Bandi, *Eskimo Prehistory*, pp. 139–142; J. T. Andrews et al., "Relative Departures in July Temperatures in Northern Canada for the Past 6,000 Yr," *Nature* 289 (1981): 164–167; Maxwell, *Prehistory of the Eastern Arctic*, pp. 129–159.

40. William E. Taylor and George Swinton, "Prehistoric Dorset Art," *The Beaver* (winter 1967): 32–47; McGhee, *Ancient People of the Arctic*, pp. 149–173.

41. Maxwell, *Prehistory of the Eastern Arctic*, pp. 228–232.

42. Peter Schledermann, "Preliminary Results of Archaeological Investigations in the Bache Peninsula Region, Ellesmere Island, N.W.T.," *Arctic* 31, no. 4 (1978): 459–474; Maxwell, *Prehistory of the Eastern Arctic*, pp. 232–233; Maribeth S. Murray, "Local Heroes: The Long-Term Effects of Short-Term Prosperity—an Example from the Canadian Arctic," *World Archaeology* 30, no. 3 (1999): 466–483; John F.

Hoffecker, *Desolate Landscapes: Ice-Age Settlement of Eastern Europe* (New Brunswick, NJ: Rutgers University Press, 2002), pp. 244–246.

43. Kaj Birket-Smith, *The Eskimos* (London: Methuen, 1959); Giddings, *Ancient Men of the Arctic,* pp. 69–71; McGhee, *Ancient People of the Arctic,* pp. 5–11.

44. Dumond, *The Eskimos and Aleuts,* pp.11–31. The English language is spoken from Great Britain to New Zealand—covering a comparable longitudinal range—but the distribution of English speakers within this range is not continuous.

45. For example, see Ackerman, "Settlements and Sea Mammal Hunting," pp. 67–68; McGhee, *Canadian Arctic Prehistory,* pp. 74–82; Don E. Dumond, "The Norton Tradition," *Arctic Anthropology* 37, no. 2 (2000): 1–22.

46. Giddings, *Ancient Men of the Arctic,* pp. 223–245; Anderson, "Prehistory of North Alaska," p. 85.

47. Dikov, *Drevnie Kul'tury Severo-Vostochnoi Azii,* pp. 165–168; Robert E. Ackerman, "Prehistory of the Asian Eskimo Zone," in *Handbook of the North American Indian,* vol. 5, *Arctic,* ed. Damas, pp. 106–118.

48. Giddings, *Ancient Men of the Arctic,* pp. 200–222; Anderson, *Onion Portage,* pp. 103–112; Dumond, "The Norton Tradition," pp. 9–13.

49. Giddings *Ancient Men of the Arctic,* pp. 175–199; John Bockstoce, *The Archaeology of Cape Nome, Alaska* (University Museum Monograph, no. 38), pp. 31–58; Dumond, "The Norton Tradition," pp. 2–6.

50. A. P. Okladnikov and N. A. Beregovaya, *Drevnie Poseleniya Baranova Mysa* (Novosibirsk: Nauka, 1971); Ackerman, "Prehistory of the Asian Eskimo Zone," pp. 108–109.

51. Henry B. Collins, *Archaeology of St. Lawrence Island, Alaska* (Smithsonian Miscellaneous Collections, vol. 96, no. 1, 1937); Froelich G. Rainey, "Eskimo Prehistory: The Okvik Site on the Punuk Islands," *Anthropological Papers of the American Museum of Natural History* 37, no. 4 (1941): 443–569; S. I. Rudenko, "The Ancient Culture of the Bering Sea and the Eskimo Problem," *Arctic Institute of North America, Anthropology of the North, Translations from Russian Sources,* no. 1 (1961).

52. Bandi, *Eskimo Prehistory,* pp. 67–81; Dumond, *The Eskimos and Aleuts,* pp. 118–125.

53. Collins, *Archaeology of St. Lawrence Island, Alaska;* Henry B. Collins, "The Arctic and Subarctic," in *Prehistoric Man in the New World,* ed. J. D. Jennings and E. Norbeck, pp. 85–114 (Chicago: University of Chicago Press, 1964), pp. 90–101; Ackerman, "Prehistory of the Asian Eskimo Zone," pp. 109–113; Owen K. Mason, "Archaeological Rorshach in Delineating Ipiutak, Punuk and Birnirk in NW Alaska: Masters, Slaves, or Partners in Trade?" in *Identities and Cultural Contacts in the Arctic,* ed. M. Appelt, J. Berglund, and H. C. Gulløv, pp. 229–251 (Copenhagen: Danish National Museum and Danish Polar Center, 2000).

54. James A. Ford, *Eskimo Prehistory in the Vicinity of Point Barrow, Alaska* (Anthropological Papers of the American Museum of Natural History, vol. 47, p. 1, 1959); Stanford, *The Walakpa Site, Alaska;* Dumond, *The Eskimos and Aleuts,* pp. 131–133.

55. Owen K. Mason, "The Contest between the Ipiutak, Old Bering Sea, and

Birnirk Polities and the Origin of Whaling during the First Millennium A.D. along Bering Strait," *Journal of Anthropological Archaeology* 17 (1998): 240–325.

56. Helge Larsen and Froelich Rainey, *Ipiutak and the Arctic Whale Hunting Culture* (Anthropological Papers of the American Museum of Natural History, vol. 42, 1948); Giddings, *Ancient Men of the Arctic*, pp. 102–150; Anderson, "Prehistory of North Alaska," pp. 88–90.

57. Larsen and Rainey, *Ipiutak and the Arctic Whale Hunting Culture*, pp. 119–146; Mason, "Contest," pp. 271–281.

58. Robert McGhee, "Speculations on Climate Change and Thule Culture Development," *Folk* 11/12 (1970): 173–184; J. T. Andrews and G. H. Miller, "Climatic Change over the Last 1000 Years, Baffin Island, N.W.T.," in *Thule Eskimo Culture: An Anthropological Retrospective*, ed. A. P. McCartney, pp. 541–554 (Ottawa: National Museums of Canada, 1979); Ackerman, "Settlements and Sea Mammal Hunting," p. 68.

59. Giddings, *Ancient Men of the Arctic*, pp. 231–232.

60. Maxwell, *Prehistory of the Eastern Arctic*, pp. 250–261.

61. McGhee, *Canadian Arctic Prehistory*, pp. 89–99. The trip alarm for hunting seals at breathing holes on the ice was described by Franz Boas, *The Central Eskimo* (Sixth Annual Report Bureau of Ethnology, Smithsonian Institution, 1888), along with other elements of technology used during the Thule period.

62. McGhee, *Canadian Arctic Prehistory*, pp. 92–97.

63. Inuit legends of the "Tunnit" people apparently refer to the Dorset and suggest some conflict. See Maxwell, *Prehistory of the Eastern Arctic*, pp. 127–128; McGhee, *Ancient People of the Arctic*.

64. Andrews and Miller, "Climatic Change over the Last 1000 Years"; W. W. Fitzhugh, *Environmental Archaeology and Cultural Systems in Hamilton Inlet, Labrador: A Survey of the Central Labrador Coast from 3000 B.C. to the Present* (Smithsonian Contributions to Anthropology, no. 16, 1972); Dumond, *The Eskimos and Aleuts*, pp. 145–149.

65. McGhee, *Canadian Arctic Prehistory*, pp. 103–117.

66. J. M. Nielsen, *Armed Forces on a Northern Frontier: The Military in Alaska's History* (New York: Greenwood Press, 1988), p. 196; S. J. Zaloga, *Target America: The Soviet Union and the Strategic Arms Race, 1945–1964* (Novato, Calif.: Presidio Press, 1993); D. Colt Denfeld, *The Cold War in Alaska: A Management Plan for Cultural Resources* (Anchorage: U.S. Army Corps of Engineers, 1994).

67. S. Duke, *U.S. Military Forces and Installations in Europe* (New York: Oxford University Press, 1989).

68. Kenneth Schaffel, *The Emerging Shield: The Air Force and the Evolution of Continental Air Defense, 1945–1960* (Washington, DC: Office of Air Force History, 1991), pp. 209–217; Maxwell, *Prehistory of the Eastern Arctic*, p. 310.

69. Robert Buderi, *The Invention That Changed the World: How a Small Group of Radar Pioneers Won the Second World War and Launched a Technological Revolution* (New York: Simon and Schuster, 1996), pp. 412–416.

Bibliography

Abramova, Z. A. "Must'erskii Grot Dvuglazka v Khakasii (Predvaritel'noe Soobshschenie)." *Kratkie Soobshcheniya Instituta Arkheologii* 165 (1981): 74–78.
———. *Paleolit Eniseya: Kokorevskaya Kul'tura.* Novosibirsk: Nauka, 1979.
———. *Paleolit Eniseya: Afontovskaya Kul'tura.* Novosibirsk: Nauka, 1979.
———. "Two Examples of Terminal Paleolithic Adaptations." In *From Kostenki to Clovis,* edited by O. Soffer and N. D. Praslov, 85–100. New York: Plenum Press, 1993.
Ackerman, Robert E. "The Neolithic-Bronze Age Cultures of Asia and the Norton Phase of Alaskan Prehistory." *Arctic Anthropology* 19, no. 2 (1982): 11–38.
———. "Prehistory of the Asian Eskimo Zone." In *Handbook of the North American Indian,* vol. 5, *Arctic,* edited by D. Damas, 106–118. Washington, DC: Smithsonian Institution, 1984.
———. "Settlements and Sea Mammal Hunting in the Bering-Chukchi Sea Region." *Arctic Anthropology* 25, no. 1 (1988): 52–79.
Aitkin, Martin J. "Chronometric Techniques for the Middle Pleistocene." In *The Earliest Occupation of Europe,* edited by W. Roebroeks and T. van Kolfschoten, 269–277. Leiden: University of Leiden, 1995.
Aksenov, M. P. "Archaeological Investigations at the Stratified Site of Verkholenskaia Gora in 1963–1965." *Arctic Anthropology* 6, no. 1 (1969): 74–87.
Alekseev, V. P. "The Physical Specificities of Paleolithic Hominids in Siberia." In *The Paleolithic of Siberia: New Discoveries and Interpretations,* edited by A. P. Derevianko, 329–335. Urbana: University of Illinois Press, 1998.
Anconetani, P. "Lo Studio Arcezoologico del Sito di Isernia La Pineta." In *I Reperti Paleontologici del Giacimento Paleolitico di Isernia La Pineta: L'uomo e L'ambiente,* edited by C. Peretto, 87–186. Isernia, 1996.
Anderson, Douglas D. *Onion Portage: The Archaeology of a Stratified Site from the Kobuk River, Northwest Alaska.* Anthropological Papers of the University of Alaska, vol. 22, nos. 1–2 (1988).
———. "Prehistory of North Alaska." In *Handbook of the North American Indian,* vol. 5, *Arctic,* edited by D. Damas, 80–93. Washington, DC: Smithsonian Institution, 1984.
———. "A Stone Age Campsite at the Gateway to America." *Scientific American* 218, no. 6 (1968): 24–33.
Anderson-Gerfaud, Patricia. "Aspects of Behaviour in the Middle Palaeolithic: Functional Analysis of Stone Tools from Southwest France." In *The Emergence of Modern Humans,* edited by P. Mellars, 389–418. Edinburgh: Edinburgh University Press, 1990.
Andrews, J. T., P. T. Davis, W. N. Mode, H. Nichols, and S. K. Short. "Relative Departures in July Temperatures in Northern Canada for the Past 6,000 Yr." *Nature* 289 (1981): 164–167.

Andrews, J. T., and G. H. Miller. "Climatic Change over the Last 1000 Years, Baffin Island, N.W.T." In *Thule Eskimo Culture: An Anthropological Retrospective*, edited by A. P. McCartney, 541–554. Ottawa: National Museums of Canada, 1979.

Andrews, Peter. "Hominoid Evolution." *Nature* 295 (1982): 185–186.

Anikovich, M. V. "The Early Upper Paleolithic in Eastern Europe." *Archaeology, Ethnology & Anthropology of Eurasia* 2, no. 14 (2003): 15–29.

Archibold, O. W. *Ecology of World Vegetation*. London: Chapman and Hall, 1995.

Arneborg, Jette, Jan Heinemeier, Niels Lynnerup, Henrik L. Nielsen, Niels Rud, and Arnd E. Svieinbjornsdottir. "Change of Diet of the Greenland Vikings Determined from Stable Carbon Isotope Analysis and 14c Dating of Their Bones." *Radiocarbon* 41, no. 2 (1999): 157–168.

Ascenzi, A., I. Biddittu, P. F. Cassoli, A. G. Segre, and E. Segre-Naldini. "A Calvarium of Late *Homo erectus* from Ceprano, Italy." *Journal of Human Evolution* 31 (1996): 409–423.

Atkinson, T. C., K. R. Briffa, and G. R. Coope, "Seasonal Temperatures in Britain during the Last 22,000 Years, Reconstructed Using Beetle Remains." *Nature* 325 (1987): 587–592.

Bader, O. N. "Pogrebeniya v Verkhnem Paleolite i Mogila na Stoyanke Sungir'." *Sovetskaya Arkheologiya* 3 (1967): 142–159.

———. *Sungir' Verkhnepaleoliticheskaya Stoyanka*. Moscow: Nauka, 1978.

———. "Vtoraya Paleoliticheskaya Mogila na Sungire." In *Arkheologicheskie Otkrytiya 1969 Goda*, 41–43. Moscow: Nauka, 1970.

Bahn, Paul G., ed. *The Cambridge Illustrated History of Archaeology*. Cambridge: Cambridge University Press, 1996.

Bahn, Paul G., and Jean Vertut. *Journey through the Ice Age*. Berkeley: University of California Press, 1997.

Bandi, Hans-Georg. *Eskimo Prehistory*. Translated by A. E. Keep. College: University of Alaska Press, 1969.

Bar-Yosef, Ofer. "The Middle and Upper Paleolithic in Southwest Asia and Neighboring Regions." In *The Geography of Neandertals and Modern Humans in Europe and the Greater Mediterranean*, edited by O. Bar-Yosef and D. Pilbeam, 107–156. Cambridge, MA: Peabody Museum of Archaeology and Ethnology, 2000.

———. "Pleistocene Connexions between Africa and Southwest Asia: An Archaeological Perspective." *African Archaeological Review* 5 (1987): 29–38.

———. "Upper Pleistocene Cultural Stratigraphy in Southwest Asia." In *The Emergence of Modern Humans*, edited by E. Trinkaus, 154–180. Cambridge: Cambridge University Press, 1989.

Bar-Yosef, Ofer, and Anna Belfer-Cohen. "From Africa to Eurasia—Early Dispersals." *Quaternary International* 75 (2001): 19–28.

Baryshnikov, Gennady, and John F. Hoffecker. "Mousterian Hunters of the NW Caucasus: Preliminary Results of Recent Investigations." *Journal of Field Archaeology* 21 (1994): 1–14.

Bermudez de Castro, J. M., J. L. Arsuaga, E. Carbonell, A. Rosas, I. Martinez, and M. Mosquera. "A Hominid from the Lower Pleistocene of Atapuerca, Spain:

Possible Ancestor to Neandertals and Modern Humans." *Science* 276 (1997): 1392–95.
Beyries, Sylvie. "Functional Variability of Lithic Sets in the Middle Paleolithic." In *Upper Pleistocene Prehistory of Western Eurasia*, edited by H. L. Dibble and A. Montet-White, 213–224. Philadelphia: University of Pennsylvania Museum, 1988.
Bibikov, S. N. *Drevneishii muzykal'nyi kompleks iz kostei mamonta*. Kiev: Naukova dumka, 1981.
Bickerton, Derek. *Language and Human Behavior*. Seattle: University of Washington Press, 1995.
———. *Language and Species*. Chicago: University of Chicago Press, 1990.
Bigelow, Nancy H., and Wm. Roger Powers. "Climate, Vegetation, and Archaeology 14,000–9000 cal yr B.P. in Central Alaska." *Arctic Anthropology* 38, no. 2 (2001): 171-195.
Binford, Lewis R. *Bones: Ancient Men and Modern Myths*. New York: Academic Press, 1981.
———. "Hard Evidence." *Discover* (February 1992): 44–51.
———. "Mobility, Housing, and Environment: A Comparative Study." *Journal of Anthropological Research* 46 (1990): 119–152.
———. "Willow Smoke and Dogs' Tails: Hunter-Gatherer Settlement Systems and Archaeological Site Formation." *American Antiquity* 45, no. 1 (1980): 4–20.
Birdsell, J. B. "The Recalibration of a Paradigm for the First Peopling of Greater Australia." In *Sunda and Sahul: Prehistoric Studies in Southeast Asia, Melanesia, and Australia*, edited by J. Allen, J. Golson, and R. Jones, 113–167. London: Academic Press, 1977.
Birket-Smith, Kaj. *The Eskimos*. London: Methuen, 1959.
Blumenschine, Robert J. "Early Hominid Scavenging Opportunities." *British Archaeological Reports International Series* 283 (1986).
Boas, Franz. *The Central Eskimo*. Sixth Annual Report of the Bureau of Ethnology, Smithsonian Institution, 1888.
Bocherens, H., D. Billiou, A. Mariotti, M. Patou-Mathis, M. Otte, D. Bonjean, and M. Toussaint. "Palaeoenvironmental and Palaeodietary Implications of Isotopic Biogeochemistry of Last Interglacial Neanderthal and Mammal Bones at Scladina Cave (Belgium)." *Journal of Archaeological Science* 26 (1999): 599–607.
Bockstoce, John. *The Archaeology of Cape Nome, Alaska*. University Museum Monograph, no. 38. University of Pennsylvania 1979.
Bocquet-Appel, Jean-Pierre, and Pierre Yves Demars. "Neanderthal Contraction and Modern Human Colonization of Europe." *Antiquity* 74 (2000): 544–552.
Boëda, E., J. Connan, D. Dessort, S. Muhesen, N. Mercier, H. Valladas, and N. Tisnerat. "Bitumen as a Hafting Material on Middle Paleolithic Artefacts." *Nature* 380 (1996): 336–338.
Bond, Gerard, Wallace Broecker, Sigfus Johnsen, Jerry McManus, Laurent Labeyrie, Jean Jouzel, and Georges Bonani. "Correlations between Climate Records from North Atlantic Sediments and Greenland Ice." *Nature* 365 (1993): 143–147.

Bond, Gerard, et al. "Evidence for Massive Discharges of Icebergs into the North Atlantic Ocean during the Last Glacial Period." *Nature* 360 (1992): 245–249.

Boorstin, Daniel J. *The Discoverers: A History of Man's Search to Know His World and Himself.* New York: Random House, 1983.

Bordes, François. *The Old Stone Age.* Translated by J. E. Anderson. New York: McGraw-Hill Book Co., 1968.

Boriskovskii, P. I. *Ocherki po Paleolitu Basseina Dona.* Materialy i Issledovaniya po Arkheologii SSSR 121, 1963.

Bosi, Roberto. *The Lapps.* London: Thames and Hudson, 1960.

Bosinski, Gerhard. "The Earliest Occupation of Europe: Western Central Europe." In *The Earliest Occupation of Europe,* edited by W. Roebroeks and T. van Kolfschoten, 103–128. Leiden: University of Leiden, 1995.

Bowler, J. M., H. Johnston, J. M. Olley, J. R. Prescott, R. G. Roberts, W. Shawcross, and N. A. Spooner. "New Ages for Human Occupation and Climatic Change at Lake Mungo, Australia." *Nature* 421 (2003): 837–840.

Brace, C. Loring. "The Fate of the 'Classic' Neanderthals: A Consideration of Hominid Catastrophism." *Current Anthropology* 5 (1964): 3–43.

Brain, C. K., and A. Sillent, "Evidence from the Swartkrans Care for the Earliest Use of Fire." *Nature* 336 (1988): 464–466.

Bronk-Ramsey, C. "Radiocarbon Calibration and Analysis of Stratigraphy: The OxCal Program." *Radiocarbon* 37 (1995): 425–430.

Buderi, Robert. *The Invention That Changed the World: How a Small Group of Radar Pioneers Won the Second World War and Launched a Technological Revolution.* New York: Simon and Schuster, 1996.

Buisson, D. "Les Flûtes Paléolithiques d'Isturitz (Pyrénées Atlantiques)." *Société Préhistorique Française* 87 (1991): 420–433.

Burov, Grigoriy M. "Some Mesolithic Wooden Artifacts from the Site of Vis I in the European North East of the U.S.S.R." In *The Mesolithic in Europe,* edited by C. Bonsall, 391–401. Edinburgh: John Donald Publishers, 1989.

Butzer, Karl W., G. J. Flock, L. Scott, and R. Stuckenrath. "Dating and Context of Rock Engravings in Southern Africa." *Science* 203 (1979): 1201–1214.

Cachel, Susan and J.W.K. Harris. "The Lifeways of *Homo erectus* Inferred from Archaeology and Evolutionary Ecology: A Perspective from East Africa." In *Early Human Behaviour in Global Context: The Rise and Diversity of the Lower Palaeolithic Record,* edited by M. D. Petraglia and R. Korisetter, 108–132. London: Routledge, 1998.

Calvin, William H., and Derek Bickerton. *Lingua ex Machina: Reconciling Darwin and Chomsky with the Human Brain.* Cambridge: MIT Press, 2000.

Carbonell, E., J. M. Bermudez de Castro, J. L. Arsuaga, J. C. Diez, A. Rosas, G. Cuenca-Bescos, R. Sala, M. Mosquera, and X. P. Rodriguez. "Lower Pleistocene Hominids and Artifacts from Atapuera-TD6 (Spain)." *Science* 269 (1995): 826–830.

Carbonell, E., and Z. Castro-Curel. "Palaeolithic Wooden Artifacts from the Abric Romani (Capellades, Barcelona, Spain)." *Journal of Archaeological Science* 19 (1992): 707–719.

Carbonell, E., M. D. García-Anton, C. Mallol, M. Mosquera, A. Olle, X. P. Rodriguez, M. Sahnouni, R. Sala, and J. M. Verges. "The TD6 Level Lithic Industry from Gran Dolina, Atapuerca (Burgos, Spain): Production and Use." *Journal of Human Evolution* 37, nos. 3–4 (1999): 653–694.

Carbonell, Eudald, and Xose Pedro Rodriguez. "Early Middle Pleistocene Deposits and Artefacts in the Gran Dolina Site (TD4) of the 'Sierra de Atapuerca' (Burgos, Spain)." *Journal of Human Evolution* 26 (1994): 291–311.

Castro-Curel, Z., and E. Carbonell. "Wood Pseudomorphs from Level I at Abric Romani, Barcelona, Spain." *Journal of Field Archaeology* 22 (1995): 376–384.

Cattelain, Pierre. "Un Crochet de Propulseur Solutréen de la Grotte de Combe-Saunière I (Dordogne)." *Bulletin de la Société Préhistorique Française* 86 (1989): 213–216.

———. "Hunting during the Upper Paleolithic: Bow, Spearthrower, or Both?" In *Projectile Technology*, edited by H. Knecht, 213–240. New York: Plenum Press, 1997.

Chard, Chester S. *Northeast Asia in Prehistory*. Madison: University of Wisconsin Press, 1974.

Chase, Philip G. "The Hunters of Combe Grenal: Approaches to Middle Paleolithic Subsistence in Europe." *British Archaeological Reports International Series* S-286, 1986.

Chase, Philip G., and Harold L. Dibble. "Middle Paleolithic Symbolism: A Review of Current Evidence and Interpretations." *Journal of Anthropological Archaeology* 6 (1987): 263–296.

Chauvet, Jean-Marie, Eliette Brunel Deschamps, and Christian Hillaire. *Dawn of Art: The Chauvet Cave*. New York: Harry N. Abrams, 1996.

Chemillier, Marc. "Ethnomusicology, Ethnomathematics: The Logic Underlying Orally Transmitted Artistic Practices." In *Mathematics and Music*, edited by G. Assayag, H. G. Feichtinger, and J. F. Rodrigues, 161–183. Berlin: Springer-Verlag, 2002.

Childe, V. Gordon. *Man Makes Himself*. London: Watts and Co., 1936.

———. *What Happened in History*. Rev ed. Harmondsworth: Penguin Books, 1954.

Churchill, Steven Emilio. "Cold Adaptation, Heterochrony, and Neandertals." *Evolutionary Anthropology* 7, no. 2 (1998): 46–61.

Clark, Grahame. *World Prehistory in New Perspective*. Illus. 3rd ed. Cambridge: University of Cambridge Press, 1977.

Clark, Grahame, and Stuart Piggot. *Prehistoric Societies*. Harmondsworth: Penguin Books, 1970.

Clark, J.G.D. "Neolithic Bows from Somerset, England, and the Prehistory of Archery in North-west Europe." *Proceedings of the Prehistoric Society* 29 (1963): 50–98.

———. *Prehistoric Europe: The Economic Basis*. Stanford: Stanford University Press, 1952.

Collins, Henry B. *Archaeology of St. Lawrence Island, Alaska*. Smithsonian Miscellaneous Collections, vol. 96, no. 1.

———. "The Arctic and Subarctic." In *Prehistoric Man in the New World*, edited by

J. D. Jennings and E. Norbeck, 85–114. Chicago: University of Chicago Press, 1964.
Conard, Nicholas J. "Palaeolithic Ivory Sculptures from Southwestern Germany and the Origins of Figurative Art." *Nature* 426 (2003): 830–832.
Conklin, Harold C. "Lexicographical Treatment of Folk Taxonomies." *International Journal of American Linguistics* 28 (1962).
Conroy, Glenn C. *Primate Evolution.* New York: W. W. Norton and Co., 1990.
Coolidge, F. L., and T. Wynn. "Executive Functions of the Frontal Lobes and the Evolutionary Ascendancy of *Homo sapiens.*" *Journal of Human Evolution* 42, no. 3 (2002): A12–A13.
Coon, Carleton S. *The Hunting Peoples.* New York: Little, Brown and Co., 1971.
———. *The Origin of Races.* New York: Alfred A. Knopf, 1962.
Coope, G. Russell. "Late-Glacial (Anglian) and Late-Temperate (Hoxnian) Coleoptera." In *The Lower Palaeolithic Site at Hoxne, England,* edited by R. Singer, B. G. Gladfelter, and J. J. Wymer, 156–162. Chicago: University of Chicago Press, 1993.
Coope, G. Russell, and Scott A. Elias. "The Environment of Upper Palaeolithic (Magdalenian and Azilian) Hunters at Hauterive-Champréveyres, Neuchâtel, Switzerland, Interpreted from Coleopteran Remains." *Journal of Quaternary Science* 15 (2000): 157–175.
Daniel, Glyn. *The Idea of Prehistory.* Harmondsworth: Penguin Books, 1962.
Dansgaard, W., S. J. Johnson, H. B. Clausen, D. Dahl-Jensen, N. S. Gundestrup, C. U. Hammer, C. S. Hvidberg, J. P. Steffensen, A. E. Sveinbjörnsdottir, J. Jouzel, and G. Bond. "Evidence for General Instability of Past Climate from a 250-kyr Ice-Core Record." *Nature* 364 (1993): 218–220.
Darwin, Charles. *The Descent of Man and Selection in Relation to Sex.* London: John Murray, 1871.
———. *On the Origin of Species.* London: John Murray, 1859.
Davidson, Iain, and William Noble. "The Archaeology of Perception: Traces of Depiction and Language." *Current Anthropology* 30, no. 2 (1989): 125–155.
Deetz, James. *Invitation to Archaeology.* Garden City, NY: Natural History Press, 1967.
Denfeld, D. Colt. *The Cold War in Alaska: A Management Plan for Cultural Resources.* Anchorage: U.S. Army Corps of Engineers, 1994.
Dennell, Robin. *European Economic Prehistory: A New Approach.* London: Academic Press, 1983.
Dennell, Robin, and Wil Roebroeks. "The Earliest Colonization of Europe: The Short Chronology Revisited." *Antiquity* 70 (1996): 535–542.
d'Errico, Francesco. "Palaeolithic Lunar Calendars: A Case of Wishful Thinking?" *Current Anthropology* 30, no. 1 (1989): 117–118.
d'Errico, Francesco, Paola Villa, Ana C. Pinto Llona, and Rosa Ruiz Idarraga. "A Middle Palaeolithic Origin of Music? Using Cave-Bear Bone Accumulations to Assess the Divje Babe I Bone 'Flute.'" *Antiquity* 72 (1998): 65–79.
de Vos, J., P. Sondaar, and C. C. Swisher. "Dating Hominid Sites in Indonesia." *Science* 266 (1994): 1726–27.

Diamond, Jared M. "Zoological Classification System of a Primitive People." *Science* 151 (1966): 1102–1104.

Díez, J. Carlos, Yolanda Fernández-Jalvo, Jordi Rosell, and Isabel Cáceres. "Zooarchaeology and Taphonomy of Aurora Stratum (Gran Dolina, Sierra de Atapuerca, Spain). *Journal of Human Evolution* 37, nos. 3–4 (1999): 623–652.

Dikov, N. N. *Arkheologicheskie Pamyatniki Kamchatki, Chukotki i Verkhnei Kolymy.* Moscow: Nauka, 1977.

———. *Drevnie Kul'tury Severo-Vostochnoi Azii.* Moscow: Nauka, 1979.

———. "The Ushki Sites, Kamchatka Peninsula." In *American Beginnings: The Prehistory and Palaeoecology of Beringia,* edited by F. H. West, 244–250. Chicago: University of Chicago Press, 1996.

Dillehay, T. D., and M. B. Collins. "Early Cultural Evidence from Monte Verde in Chile." *Nature* 332 (1988): 150–152.

Dixon, E. James. *Bones, Boats, and Bison.* Albuquerque: University of New Mexico Press, 1999.

———. "Cultural Chronology of Central Interior Alaska." *Arctic Anthropology* 22, no. 1 (1985): 47–66.

———. "Human Colonization of the Americas: Timing, Technology, and Process." *Quaternary Science Reviews* 20, nos. 1–3 (2001): 277–299.

Dolukhanov, P. M., and N. A. Khotinskiy. "Human Cultures and Natural Environment in the USSR during the Mesolithic and Neolithic." In *Late Quaternary Environments of the Soviet Union,* edited by A. A. Velichko, 319–327. Minneapolis: University of Minnesota Press, 1984.

Dolukhanov, P., D. Sokoloff, and A. Shukurov. "Radiocarbon Chronology of Upper Palaeolithic Sites in Eastern Europe at Improved Resolution." *Journal of Archaeological Science* 28 (2001): 699–712.

Douglas, Mary. "Symbolic Orders in the Use of Domestic Space." In *Man, Settlement, and Urbanism,* edited by P. J. Ucko, R. Tringham, and G. W. Dimbleby, 513–521. Cambridge, MA: Schenkman Publishing Co., 1970.

Duarte, C., J. Mauricio, P. B. Pettitt, P. Souto, E. Trinkaus, H. van der Plicht, and J. Zilhao. "The Early Upper Paleolithic Human Skeleton from the Abrigo do Lagar Velho (Portugal) and Modern Human Emergence in Iberia." *Proceedings of the National Academy of Sciences* 96 (1999): 7604–7609.

Duke, S. *U.S. Military Forces and Installations in Europe.* New York: Oxford University Press, 1989.

Dumond, Don E. *The Eskimos and Aleuts.* Rev. ed. London: Thames and Hudson, 1987.

———. "The Norton Tradition." *Arctic Anthropology* 37, no. 2 (2000): 1–22.

Efimenko, P. P. *Kostenki I.* Moscow: USSR Academy of Sciences, 1958.

Eldredge, Niles. *Time Frames: The Evolution of Punctuated Equilibria.* Princeton: Princeton University Press, 1985.

Elias, Scott A. "Beringian Paleoecology: Results from the 1997 Workshop." *Quaternary Science Reviews* 20, no. 1 (2001): 7–13.

———. *Quaternary Insects and Their Environments*. Washington, DC: Smithsonian Institution Press, 1994.

Elias, Scott A., Susan K. Short, and R. Lawrence Phillips. "Paleoecology of Late-Glacial Peats from the Bering Land Bridge, Chukchi Sea Shelf Region, Northwestern Alaska." *Quaternary Research* 38 (1992): 371–378.

Enard, W., et al. "Intra- and Interspecific Variation in Primate Gene Expression Patterns." *Science* 296 (2002): 340–343.

Enard, Wolfgang, Molly Przeworski, Simon E. Fisher, Cecelia S. L. Lai, Victor Wiebe, Takashi Kitano, Anthony P. Monaco, and Svante Pääbo. "Molecular Evolution of FOXP2, a Gene Involved in Speech and Language." *Nature* 418 (2002): 869–872.

Engelstad, Ericka. "The Late Stone Age of Arctic Norway: A Review." *Arctic Anthropology* 22, no. 1 (1985): 79–96.

———. "Mesolithic House Sites in Arctic Norway." In *The Mesolithic in Europe*, edited by C. Bonsall, 331–337. Edinburgh: John Donald Publishers, 1989.

Ermolova, N. M. *Teriofauna Doliny Angary v Pozdnem Antropogene*. Novosibirsk: Nauka, 1978.

Fagan, Brian M. *The Journey from Eden: The Peopling of Our World*. London: Thames and Hudson, 1990.

———. *The Little Ice Age: How Climate Made History, 1300–1850*. New York: Basic Books, 2000.

Falk, Dean. "Hominid Brain Evolution and the Origins of Music." In *The Origins of Music*, edited by N. L. Wallin, B. Merker, and S. Brown, 197–216. Cambridge: MIT Press, 2000.

Feathers, James K. "Luminescence Dating and Modern Human Origins." *Evolutionary Anthropology* 5, no. 1 (1996): 25–36.

Féblot-Augustins, J. "Raw Material Transport Patterns and Settlement Systems in the European Lower and Middle Palaeolithic: Continuity, Change, and Variability." In *The Middle Palaeolithic Occupation of Europe*, edited by W. Roebroeks and C. Gamble, 193–214. Leiden: University of Leiden, 1999.

Fernández-Jalvo, Yolanda, J. Carlos Díez, Isabel Cáceres, and Jordi Rosell. "Human Cannibalism in the Early Pleistocene of Europe (Gran Dolina, Sierra de Atapuerca, Burgos, Spain)." *Journal of Human Evolution* 37, nos. 3–4 (1999): 591–622.

Fitzhugh, William W. "A Comparative Approach to Northern Maritime Adaptations." In *Prehistoric Maritime Adaptations of the Circumpolar Zone*, edited by W. Fitzhugh, 339–386. The Hague: Mouton Publishers, 1975.

———. *Environmental Archaeology and Cultural Systems in Hamilton Inlet, Labrador: A Survey of the Central Labrador Coast from 3000 B.C. to the Present*. Smithsonian Contributions to Anthropology, no. 16. 1972.

Fitzhugh, William W., and Elisabeth I. Ward, eds. *Vikings: The North Atlantic Saga*. Washington, DC: Smithsonian Institution Press, 2000.

Foley, Robert A. "The Evolution of Hominid Social Behaviour." In *Comparative Socioecology: The Behavioural Ecology of Humans and Other Mammals*, edited by

V. Standen and R. A. Foley, 473–494. Oxford: Blackwell Scientific Publications, 1989.
Ford, James A. *Eskimo Prehistory in the Vicinity of Point Barrow, Alaska.* Anthropological Papers of the American Museum of Natural History, vol. 47, pt. 1. 1959.
Franciscus, Robert G., and Steven E. Churchill. "The Costal Skeleton of Shanidar 3 and a Reappraisal of Neandertal Thoracic Morphology." *Journal of Human Evolution* 42 (2002): 303–356.
Franciscus, Robert G., and Erik Trinkaus. "Nasal Morphology and the Emergence of *Homo erectus.*" *American Journal of Physical Anthropology* 75, no. 4 (1988): 517–527.
Freeman, Leslie G. "Acheulean Sites and Stratigraphy in Iberia and the Maghreb." In *After the Australopithecines,* edited by K. W. Butzer and G. Ll. Isaac, 661–743. The Hague: Mouton Publishers, 1975.
Gabunia, Leo, et al. "Dmanisi and Dispersal." *Evolutionary Anthropology* 10 (2001): 158–170.
———. "Earliest Pleistocene Cranial Remains from Dmanisi, Republic of Georgia: Taxonomy, Geological Setting, and Age." *Science* 288 (2000): 1019–1025.
Gamble, Clive. "The Earliest Occupation of Europe: The Environmental Background." In *The Earliest Occupation of Europe,* edited by W. Roebroeks and T. van Kolfschoten, 279–295. Leiden: University of Leiden, 1995.
———. *The Palaeolithic Settlement of Europe.* Cambridge: University of Cambridge Press, 1986.
———. *The Palaeolithic Societies of Europe.* Cambridge: Cambridge University Press, 1999.
———. *Timewalkers: The Prehistory of Global Colonization.* Cambridge: Harvard University Press, 1994.
Gaudzinski, Sabine. "On Bovid Assemblages and Their Consequences for the Knowledge of Subsistence Patterns in the Middle Palaeolithic." *Proceedings of the Prehistoric Society* 62 (1996): 19–39.
Gaudzinski, Sabine, and Elaine Turner. "The Role of Early Humans in the Accumulation of European Lower and Middle Palaeolithic Bone Assemblages." *Current Anthropology* 37 (1996): 153–156.
Gening, V. F., and V. T. Petrin. *Pozdnepaleolitcheskaya Epokha na Yuge Zapadnoi Sibiri.* Novosibirsk: Nauka, 1985.
Giddings, J. Louis. *Ancient Men of the Arctic.* New York: Alfred A. Knopf, 1967.
———. *The Archeology of Cape Denbigh.* Providence, RI: Brown University Press, 1964.
Gjessing, Gutorm. "Circumpolar Stone Age." *Acta Arctica* 2 (1944): 1–70.
———. "The Circumpolar Stone Age." *Antiquity* 27 (1953): 131–136.
———. "Maritime Adaptations in Northern Norway's Prehistory." In *Prehistoric Maritime Adaptations of the Circumpolar Zone,* edited by W. Fitzhugh, 87–100. The Hague: Mouton Publishers, 1975.
Gladkih, M. I., N. L. Kornietz, and O. Soffer. "Mammoth-Bone Dwellings on the Russian Plain." *Scientific American* 251, no. 5 (1984): 164–175.

Goebel, Ted. "The 'Microblade Adaptation' and Recolonization of Siberia during the Late Upper Pleistocene." In *Thinking Small: Global Perspectives on Microlithization*, edited by R. G. Elston and S. L. Kuhn, 117–131. Archaeological Papers of the American Anthropological Association no. 12. 2002.

———. "The Pleistocene Colonization of Siberia and Peopling of the Americas: An Ecological Approach." *Evolutionary Anthropology* 8 (1999): 208–227.

Goebel, Ted, A. P. Derevianko, and V. T. Petrin. "Dating the Middle-to-Upper-Paleolithic Transition at Kara-Bom." *Current Anthropology* 34 (1993): 452–458.

Goebel, Ted, and Sergei B. Slobodin. "The Colonization of Western Beringia: Technology, Ecology, and Adaptation." In *Ice Age Peoples of North America: Environments, Origins, and Adaptations of the First Americans*, edited by R. Bonnichsen and K. L. Turnmire, 104–155. Corvallis: Oregon State University Press, 1999.

Goebel, Ted, Michael R. Waters, I. Buvit, M. V. Konstantinov, and A. V. Konstantinov. "Studenoe-2 and the Origins of Microblade Technologies in the Transbaikal, Siberia." *Antiquity* 74 (2000): 567–575.

Goebel, Ted, Michael R. Waters, and Margarita Dikova. 2003. "The Archaeology of Ushki Lake, Kamchatka, and the Pleistocene Peopling of the Americas." *Science* 301 (2003): 501–505.

Goldberg, Elkhonnon. *The Executive Brain: Frontal Lobes and the Civilized Mind.* Oxford: Oxford University Press, 2001.

Goldberg, P., S. Weiner, O. Bar-Yosef, Q. Xu, and J. Liu. 2001. "Site Formation Processes at Zhoukoudian, China." *Journal of Human Evolution* 41 (2001): 483–530.

Golovanova, L. V., John F. Hoffecker, V. M. Kharitonov, and G. P. Romanova. 1999. "Mezmaiskaya Cave: A Neanderthal Occupation in the Northern Caucasus." *Current Anthropology* 41 (1999): 77–86.

Goodall, Jane. "My Life among the Wild Chimpanzees." *National Geographic Magazine* 124 (1963): 272–308.

Grichuk, V. P. "Late Pleistocene Vegetation History." In *Late Quaternary Environments of the Soviet Union*, edited by A. A. Velichko, 155–178. Minneapolis: University of Minnesota Press, 1984.

Grigor'ev, G. P. "The Kostenki-Avdeevo Archaeological Culture and the Willendorf-Pavlov-Kostenki-Avdeevo Cultural Unity." In *From Kostenki to Clovis*, edited by O. Soffer and N. D. Praslov, 51–65. New York: Plenum Press, 1993.

Grootes, P. M., M. Stuiver, J.W.C. White, S. Johnsen, and J. Jouzel. "Comparison of Oxygen Isotope Records from the GISP2 and GRIP Greenland Ice Cores." *Nature* 366 (1993): 552–554.

Gryaznov, M. P. "Ostatki Cheloveka iz Kul'turnogo Sloya Afontova Gory." *Trudy Komissii po Izucheniyu Chetverticnogo Perioda* 1 (1932): 137–144.

Gulløv, Hans Christian. "Natives and Norse in Greenland." In *Vikings: The North Atlantic Saga*, edited by W. W. Fitzhugh and E. I. Ward, pp. 318–326. Washington, DC: Smithsonian Institution Press, 2000.

Gurina, N. N. "Mezolit Karelii." In *Mezolit SSSR*, edited by L. V. Kol'tsov, 27–31. Moscow: Nauka, 1989.

———. "Mezolit Kol'skogo Poluostrova." In *Mezolit SSSR*, edited by L. V. Kol'tsov, 20–26. Moscow: Nauka, 1989.

———. "O Nekotorykh Obshchikh Elementakh Kul'tury Drevnikh Plemen Kol'skogo Poluostrova i ikh Sosedei." In *Paleolit i Neolit*, edited by V. P. Liubin, 83–92. Leningrad: Nauka, 1986.

———. "Neolit Lesnoi i Lesostepnoi Zon Evropeiskoi Chasti SSSR." In *Kamennyi Vek na Territorii SSSR*, edited by A. A. Formozov, 134–156. Moscow: Nauka, 1970.

Guthrie, R. Dale. *Frozen Fauna of the Mammoth Steppe: The Story of Blue Babe*. Chicago: University of Chicago Press, 1990.

———. "Mammals of the Mammoth Steppe as Paleoenvironmental Indicators." In *Paleoecology of Beringia*, edited by D. M. Hopkins, J. V. Matthews, C. E. Schweger, and S. B. Young, 307–326. New York: Academic Press, 1982.

———. "Origin and Causes of the Mammoth Steppe: A Story of Cloud Cover, Woolly Mammoth Tooth Pits, Buckles, and Inside-Out Beringia." *Quaternary Science Reviews* 20, nos. 1–3 (2001): 549–574.

———. "Paleoecology of the Large-Mammal Community in Interior Alaska during the Late Pleistocene." *American Midland Naturalist* 79, no. 2 (1968): 346–363.

Hagen, Anders. *Norway*. London: Thames and Hudson, 1967.

Hahn, Joachim. "Le Paléolithique Supérieur en Allemagne Méridonale (1991–1995)." *ERAUL* 76 (1996): 181–186.

Hall, Edward T. *The Dance of Life: The Other Dimension of Time*. New York: Doubleday and Co., 1983.

———. *The Hidden Dimension*. Garden City, NY: Doubleday and Co., 1966.

Harris, Marvin. *The Rise of Anthropological Theory: A History of Theories of Culture*. New York: Thomas Y. Crowell Co., 1968.

Harritt, Roger K. "Paleo-Eskimo Beginnings in North America: A New Discovery at Kuzitrin Lake, Alaska." *Etudes/Inuit/Studies* 22, no. 1 (1998): 61–81.

Harrold, Francis B. "A Comparative Analysis of Eurasian Palaeolithic Burials." *World Archaeology* 12, no. 2 (1980): 195–211.

Hart, J. S., H. B. Sabean, J. A. Hildes, F. Depocas, H. T. Hammel, K. L. Andersen, L. Irving, and G. Foy. "Thermal and Metabolic Responses of Coastal Eskimo during a Cold Night." *Journal of Applied Physiology* 17 (1962): 953–960.

Hatt, Gudmund. "Arctic Skin Clothing in Eurasia and America: An Ethnographic Study." *Arctic Anthropology* 5, no. 2 (1969): 3–132.

Helskog, Ericka. "The Komsa Culture: Past and Present." *Arctic Anthropology* 11, suppl. (1974): 261–265.

Helskog, Knut. "Boats and Meaning: A Study of Change and Continuity in the Alta Fjord, Arctic Norway, from 4200 to 500 Years B.C." *Journal of Anthropological Archaeology* 4 (1985): 177–205.

Henderson-Sellers, Ann, and Peter J. Robinson. *Contemporary Climatology*. Edinburgh Gate: Addison Wesley Longman, 1986.

Henshilwood, Christopher S., Francesco D'Errico, Curtis W. Marean, Richard G. Milo, and Royden Yates. "An Early Bone Tool Industry from the Middle Stone Age at Blombos Cave, South Africa: Implications for the Origins of Modern Human Behaviour, Symbolism, and Language." *Journal of Human Evolution* 41, no. 6 (2001): 631–678.

Hewes, Gordon W. "A History of Speculation on the Relation between Tools and Language." In *Tools, Language, and Cognition in Human Evolution*, edited by K. R. Gibson and T. Ingold, 20–31. Cambridge: Cambridge University Press, 1993.

Hoffecker, John F. *Desolate Landscapes: Ice-Age Settlement of Eastern Europe*. New Brunswick, NJ: Rutgers University Press, 2002.

———. "The Eastern Gravettian 'Kostenki Culture' as an Arctic Adaptation." *Anthropological Papers of the University of Alaska*, n.s. 2, no. 1 (2002): 115–136.

———. "Late Pleistocene and Early Holocene Sites in the Nenana River Valley, Central Alaska." *Arctic Anthropology* 38, no. 2 (2001): 139–153.

Hoffecker, J. F., M. V. Anikovich, A. A. Sinitsyn, V. T. Holliday, and S. L. Forman. "Initial Upper Paleolithic in Eastern Europe: New Research at Kostenki." *Journal of Human Evolution* 42, no. 3 (2002): A16–A17.

Hoffecker, John F., G. F. Baryshnikov, and V. B. Doronichev. "Large Mammal Taphonomy of the Middle Pleistocene Hominid Occupation at Treugol'naya Cave (Northern Caucasus)." *Quaternary Science Reviews* 22, nos. 5–7 (2003): 595–607.

Hoffecker, John F., and Naomi Cleghorn. "Mousterian Hunting Patterns in the Northwestern Caucasus and the Ecology of the Neanderthals." *International Journal of Osteoarchaeology* 10 (2000): 368–378.

Hoffecker, John F., and Scott A. Elias. "Environment and Archeology in Beringia." *Evolutionary Anthropology* 12, no. 1 (2003): 34–49.

Hoffecker, John F., W. Roger Powers, and Ted Goebel. "The Colonization of Beringia and the Peopling of the New World." *Science* 259 (1993): 46–53.

Holliday, T. W. "Brachial and Crural Indices of European Late Upper Paleolithic and Mesolithic Humans." *Journal of Human Evolution* 36 (1999): 549–566.

———. "Postcranial Evidence of Cold Adaptation in European Neandertals." *American Journal of Physical Anthropology* 104 (1997): 245–258.

Holloway, Ralph L. "The Poor Brain of *Homo sapiens neanderthalensis:* See What You Please . . ." In *Ancestors: The Hard Evidence*, edited by E. Delson, 319–324. New York: Alan R. Liss, 1985.

Holmes, Charles E. "Broken Mammoth." In *American Beginnings: The Prehistory and Palaeoecology of Beringia*, edited by F. H. West, 312–318. Chicago: University of Chicago Press, 1996.

Holmes, Charles E., Richard VanderHoek, and Thomas E. Dilley. "Swan Point." In *American Beginnings: The Prehistory and Palaeoecology of Beringia*, edited by F. H. West, 319–323. Chicago: University of Chicago Press, 1996.

Hou, Yamei, Richard Potts, Yuan Baoyin, Guo Zhengtang, Alan Deino, Wang Wei, Jennifer Clark, Xie Guangmao, and Huang Weiwen. "Mid-Pleistocene Acheulean-like Stone Technology of the Bose Basin, South China." *Science* 287 (2000): 1622–26.

Hublin, J.-J. "Climatic Changes, Paleogeography, and the Evolution of the Neandertals." In *Neandertals and Modern Humans in Western Asia*, edited by T. Akazawa, K. Aoki, and O. Bar-Yosef, 295–310. New York: Plenum Press, 1998.

Hughen, K., S. Lehman, J. Southon, J. Overpeck, O. Marchal, C. Herring, and

J. Turnbull. "¹⁴C Activity and Global Carbon Cycle Changes over the Past 50,000 Years." *Science* 303 (2004): 202–207.

Hultén, Eric. *Outline of the History of Arctic and Boreal Biota during the Quaternary Period.* Stockholm: Bokförlags Aktiebolaget Thule, 1937.

Hylander, W. L. "The Adaptive Significance of Eskimo Cranio-Facial Morphology." In *Oro-Facial Growth and Development,* edited by A. A. Dahlberg and T. Graber. The Hague: Mouton, 1977.

Ingman, M., H. Kaessmann, S. Pääbo, and U. Gyllensten. "Mitochondrial Genome Variation and the Origin of Modern Humans." *Nature* 408 (2000): 708–713.

Issac, Glynn LI. "The Archaeology of Human Origins." *Advances in World Archaeology* 3 (1984): 1–87.

———. "The Food-Sharing Behavior of Protohuman Hominids." *Scientific American* 238, no. 4 (1978): 90–108.

———. "Stages of Cultural Elaboration in the Pleistocene: Possible Archaeological Indicators of the Development of Language Capabilities." *Annals of the New York Academy of Sciences* 280 (1976): 275–288.

Jacobs, Kenneth H. "Climate and the Hominid Postcranial Skeleton in Wurm and Early Holocene Europe." *Current Anthropology* 26 (1985): 512–514.

James, Steven R. "Hominid Use of Fire in the Lower and Middle Pleistocene." *Current Anthropology* 30, no. 1 (1989): 1–26.

Jochim, Michael. "The Upper Palaeolithic." In *European Prehistory: A Survey,* edited by S. Milisauskas, 55–113. New York: Kluwer Academic/Plenum Publishers, 2002.

Jouzel, J., et al. "Extending the Vostok Ice-Core Record of Palaeoclimate to the Penultimate Glacial Period." *Nature* 364 (1993): 407–412.

Juel Jensen, H. "Functional Analysis of Prehistoric Flint Tools by High-Powered Microscopy: A Review of West European Research." *Journal of World Prehistory* 2, no. 1 (1988): 53–88.

Kaemmer, John E. *Music in Human Life: Anthropological Perspectives on Music.* Austin: University of Texas Press, 1993.

Keeley, Lawrence H. *Experimental Determination of Stone Tool Uses: A Microwear Analysis.* Chicago: University of Chicago Press, 1980.

———. "Microwear Analysis of Lithics." In *The Lower Paleolithic Site at Hoxne, England,* edited by R. Singer, B. G. Gladfelter, and J. J. Wymer, 129–138. Chicago: University of Chicago Press, 1993.

Keeley, Lawrence H., and Nicolas Toth. "Microwear Polishes on Early Stone Tools from Koobi Fora, Kenya." *Nature* 293 (1981): 464–465.

Kelly, Robert L. *The Foraging Spectrum: Diversity in Hunter-Gatherer Lifeways.* Washington, DC: Smithsonian Institution Press, 1995.

Kennedy, G. E. "The Emergence of *Homo sapiens:* The Post-cranial Evidence." *Man* 19 (1984): 94–110.

Khlobystin, L. P. "O Drevnem Zaselenii Arktiki." *Kratkie Soobshcheniya Instituta Arkheologii* 36 (1973): 11–16.

Khlobystin, L. P., and G. M. Levkovskaya. "Rol' Sotsial'nogo i Ekologicheskogo

Faktorov v Razvitii Arkticheskikh Kul'tur Evrazii." In *Pervobytnyi Chelovek, Ego Material'naya Kul'tura i Prirodnaya Sreda v Pleistotsene i Golotsene,* edited by I. P. Gerasimov, 235–242. Moscow: USSR Academy of Sciences, 1974.

Khotinskiy, N. A. "Holocene Climatic Change." In *Late Quaternary Environments of the Soviet Union,* edited by A. A. Velichko, 305–309. Minneapolis: University of Minnesota Press, 1984.

Kittler, Ralf, Manfred Kayser, and Mark Stoneking, "Molecular Evolution of *Pediculus humanus* and the Origin of Clothing." *Current Biology* 13 (2003): 1414–1417.

Kir'yak, M. A. *Arkheologiya Zapadnoi Chukotki.* Moscow: Nauka, 1993.

Klein, Richard G. "Archeology and the Evolution of Human Behavior." *Evolutionary Anthropology* 9 (2000): 17–36.

———. *The Human Career.* 2nd ed. Chicago: University of Chicago Press, 1999.

———. *Ice-Age Hunters of the Ukraine.* Chicago: University of Chicago Press, 1973.

———. "Problems and Prospects in Understanding How Early People Exploited Animals." In *The Evolution of Human Hunting,* edited by M. H. Nitecki and D. V. Nitecki, 11–45. New York: Plenum Press, 1987.

———. "Whither the Neanderthals?" *Science* 299 (2003): 1525–27.

Knuth, Eigil. "Archaeology of the Musk-Ox Way." *Contributions du Centre d'Etudes Arctiques et Finno-Scandinaves* 5 (1967).

Koc Karpuz, N., and E. Jansen. "A High-Resolution Diatom Record of the Last Deglaciation from the SE Norwegian Sea: Documentation of Rapid Climatic Changes." *Paleooceanography* 7 (1992): 499–520.

Koenigswald, W. von. "Various Aspects of Migrations in Terrestrial Animals in Relation to Pleistocene Faunas of Central Europe." *Courier Forschungsinstitut Senckenberg* 153 (1992): 39–47.

Kol'tsov, L. V. *Final'nyi Paleolit i Mezolit Yuzhnoi i Vostochnoi Pribaltiki.* Moscow: Nauka, 1977.

———. "Mezolit Severa Sibiri i Dal'nego Vostoka." In *Mezolit SSSR,* edited by L. V. Kol'tsov, 187–194. Moscow: Nauka, 1989.

Kordos, Laszlo, and David R. Begun. "Rudabanya: A Late Miocene Subtropical Swamp Deposit with Evidence of the Origin of the African Apes and Humans." *Evolutionary Anthropology* 11 (2002): 45–57.

Korniets, N. L., M. I. Gladkikh, A. A. Velichko, G. V. Antonova, Yu. N. Gribchenko, E. M. Zelikson, E. I. Kurenkova, T. Kh. Khalcheva, and A. L. Chepalyga, "Mezhirich." In *Arkheologiya i Paleogeografiya Pozdnego Paleolita Russkoi Ravniny,* edited by I. P. Gerasimov, 106–119. Moscow: Nauka, 1981.

Kozlowski, J. K. "The Gravettian in Central and Eastern Europe." In *Advances in World Archaeology,* vol. 5, edited by F. Wendorf and A. E. Close, 131–200. Orlando, FL: Academic Press, 1986.

Kozlowski, Janusz, and H.-G. Bandi. "The Paleohistory of Circumpolar Arctic Colonization." *Arctic* 37, no. 4 (1984): 359–372.

Kraatz, Reinhart. "A Review of Recent Research on Heidelberg Man, *Homo erectus heidelbergensis.*" In *Ancestors: The Hard Evidence,* edited by E. Delson, 268–271. New York: Alan R. Liss, 1985.

Kretzoi, M., and Dobosi, V., eds. *Vértesszöllös: Man, Site, and Culture.* Budapest: Akademiai Kiado, 1990.

Krings, M., C. Capelli, F. Tschentscher, H. Geisert, S. Meyer, A. von Haeseler, K. Grossschmidt, G. Possnert, M. Paunovic, and S. Pääbo. "A View of Neandertal Genetic Diversity." *Nature Genetics* 26, no. 2 (2000): 144–146.

Krings, M., A. Stone, R. W. Schmitz, H. Krainitzki, M. Stoneking, and S. Pääbo. "Neanderthal DNA Sequences and the Origin of Modern Humans." *Cell* 90 (1997): 19–30.

Kroeber, A. L. *Anthropology.* New York: Harcourt, Brace and World, 1948.

Lahr, M. M., and Foley, R. "Multiple Dispersals and Modern Human Origins." *Evolutionary Anthropology* 3 (1994): 48–60.

Lamb, Hubert H. *Climate, History, and the Modern World.* 2nd ed. London: Routledge, 1995.

Larsen, Helge, and Froelich Rainey. *Ipiutak and the Arctic Whale Hunting Culture.* Anthropological Papers of the American Museum of Natural History, vol. 42 (1948).

Laughlin, William S. "Aleuts: Ecosystem, Holocene History, and Siberian Origin." *Science* 189 (1975): 507–515.

Laville, Henri, Jean-Philippe Rigaud, and James Sackett. *Rock Shelters of the Perigord.* New York: Academic Press, 1980.

Leakey, L.S.B., P. V. Tobias, and J. R. Napier. "A New Species of the Genus *Homo* from Olduvai Gorge, Tanzania." *Nature* 202 (1964): 7–9.

Leakey, Mary D. *Olduvai Gorge: Excavations in Beds I and II, 1960–1963.* Cambridge: Cambridge University Press, 1971.

Levin, M. G., and L. P. Potapov, eds. *The Peoples of Siberia.* Translated by Stephen Dunn. Chicago: University of Chicago Press, 1964.

Lévi-Strauss, Claude. *The Savage Mind.* Chicago: University of Chicago Press, 1966.

———. *Structural Anthropology.* Translated by C. Jacobson and B. Schoepf. New York: Basic Books, 1963.

Lewin, Roger. *Bones of Contention: Controversies in the Search for Human Origins.* New York: Simon and Schuster, 1987.

———. *Human Evolution: An Illustrated Introduction.* 3rd ed. Boston: Blackwell Scientific Publications, 1993.

Lewis-Williams, David. *The Mind in the Cave: Consciousness and the Origins of Art.* London: Thames and Hudson, 2002.

Lieberman, Philip. *Eve Spoke: Human Language and Human Evolution.* New York: W. W. Norton and Co., 1998.

Lieberman, P., J. T. Laitman, J. S. Reidenberg, and P. J. Gannon. "The Anatomy, Physiology, Acoustics, and Perception of Speech: Essential Elements in the Analysis of the Evolution of Human Speech." *Journal of Human Evolution* 23 (1992): 447–467.

Lindly, John M., and Geoffrey A. Clark. "Symbolism and Modern Human Origins." *Current Anthropology* 31 (1990): 233–261.

Lisitsyn, N. F., and Yu. S. Svezhentsev. "Radiouglerodnaya Khronologiya Verkhnego

Paleolita Severnoi Azii." In *Radiouglerodnaya Khronologiya Paleolita Vostochnoi Evropy i Severnoi Azii: Problemy i Perspektivy*, edited by A. A. Sinitsyn and N. D. Praslov, 67–108. Saint Petersburg: Russian Academy of Sciences, 1997.

Lowe, J. J., and M.J.C. Walker. *Reconstructing Quaternary Environments*. 2nd ed. London: Longman, 1997.

Lydolph, Paul. *Geography of the U.S.S.R.* 3rd ed. New York: John Wiley, 1977.

Lynnerup, Niels. "Life and Death in Norse Greenland." In *Vikings: The North Atlantic Saga*, edited by W. W. Fitzhugh and E. I. Ward, 285–294. Washington, DC: Smithsonian Institution Press, 2000.

Madella, Marco, Martin K. Jones, Paul Goldberg, Yuval Goren, and Erella Hovers. "The Exploitation of Plant Resources by Neanderthals in Amud Cave (Israel): The Evidence from Phytolith Studies." *Journal of Archaeological Science* 29 (2002): 703–719.

Mandryk, Carole A. S., Heiner Josenhans, Daryl W. Fedje, and Rolf W. Mathewes. "Late Quaternary Paleoenvironments of Northwestern North America: Implications for Inland Versus Coastal Migration Routes." *Quaternary Science Reviews* 20, nos. 1–3 (2001): 301–314.

Mania, Dietrich. "The Earliest Occupation of Europe: The Elbe-Saale Region (Germany)." In *The Earliest Occupation of Europe*, edited by W. Roebroeks and T. van Kolfschoten, 85–101. Leiden: University of Leiden, 1995.

Marean, Curtis W., and Zelalem Assefa. "Zooarchaeological Evidence for the Faunal Exploitation Behavior of Neandertals and Early Modern Humans." *Evolutionary Anthropology* 8, no. 1 (1999): 22–37.

Marks, J., C. W. Schmid, and V. M. Sarich. "DNA Hybridization as a Guide to Phylogeny: Relations of the Hominoidea." *Journal of Human Evolution* 17 (1988): 769–786.

Marshack, Alexander. *The Roots of Civilization*. London: Weidenfeld and Nicolson, 1972.

———. "The Taï Plaque and Calendrical Notation in the Upper Palaeolithic." *Cambridge Archaeological Journal* 1 (1991): 25–61.

———. "Upper Paleolithic Symbol Systems of the Russian Plain: Cognitive and Comparative Analysis." *Current Anthropology* 20 (1979): 271–311.

Martin, Lawrence. "Significance of Enamel Thickness in Hominid Evolution." *Nature* 314 (1985): 260–263.

Mason, Owen K. "Archaeological Rorshach in Delineating Ipiutak, Punuk, and Birnirk in NW Alaska: Masters, Slaves, or Partners in Trade?" In *Identities and Cultural Contacts in the Arctic*, edited by M. Appelt, J. Berglund, and H. C. Gullov, 229–251. Copenhagen: Danish National Museum and Danish Polar Center, 2000.

———. "The Contest between the Ipiutak, Old Bering Sea, and Birnirk Polities and the Origin of Whaling during the First Millennium A.D. along Bering Strait." *Journal of Anthropological Archaeology* 17 (1998): 240–325.

Mason, Owen K., Peter M. Bowers, and David M. Hopkins. "The Early Holocene Milankovitch Thermal Maximum and Humans: Adverse Conditions for the De-

nali Complex of Eastern Beringia." *Quaternary Science Reviews* 20, nos. 1–3 (2001): 525–548.
Mason, Owen K., and S. Craig Gerlach. "Chukchi Hot Spots, Paleo-Polynas, and Caribou Crashes: Climatic and Ecological Dimensions of North Alaska Prehistory." *Arctic Anthropology* 32, no. 1 (1995): 101–130.
Matthews, J. V. "East Beringia during Late Wisconsin Time: A Review of the Biotic Evidence." In *Paleoecology of Beringia,* edited by D. M. Hopkins, J. V. Matthews, Jr., C. E. Schweger, and S. B. Young, 127–150. New York: Academic Press, 1982.
Maxwell, Moreau S. *Prehistory of the Eastern Arctic.* Orlando, FL: Academic Press, 1985.
McBrearty, S., and A. S. Brooks. "The Revolution That Wasn't: A New Interpretation of the Origin of Modern Human behavior." *Journal of Human Evolution* 39, no. 5 (2000): 453–563.
McBurney, Charles B. M. "The Geographical Study of the Older Palaeolithic Stages in Europe." *Proceedings of the Prehistoric Society* 16 (1950): 163–183.
McCartney, Allen P., and Douglas W. Veltre. "Anangula Core and Blade Site." In *American Beginnings: The Prehistory and Palaeoecology of Beringia,* edited by F. H. West, 443–450. Chicago: University of Chicago Press, 1996.
McGhee, Robert. *Ancient People of the Arctic.* Vancouver: University of British Columbia, 1996.
———. *Canadian Arctic Prehistory.* Hull, Quebec: Canadian Museum of Civilization, 1990.
———. "Speculations on Climate Change and Thule Culture Development." *Folk* 11/12 (1970): 173–184.
McGovern, Thomas H. "The Demise of Norse Greenland." In *Vikings: The North Atlantic Saga,* edited by W. W. Fitzhugh and E. I. Ward, 327–339. Washington, DC: Smithsonian Institution Press, 2000.
McGrew, William C. *Chimpanzee Material Culture.* Cambridge: Cambridge University Press, 1992.
Medvedev, G. I. "Results of the Investigations of the Mesolithic in the Stratified Settlement of Ust-Belaia 1957–1964." *Arctic Anthropology* 6, no. 1 (1969): 61–73.
———. "Upper Paleolithic Sites in South-Central Siberia." In *The Palaeolithic of Siberia: New Discoveries and Interpretations,* edited by A. P. Derevianko, 122–132. Urbana: University of Illinois Press, 1998.
Mellars, Paul. "Major Issues in the Emergence of Modern Humans." *Current Anthropology* 30, no. 3 (1989): 349–385.
———. *The Neanderthal Legacy: An Archaeological Perspective from Western Europe.* Princeton: Princeton University Press, 1996.
Merriam, Alan P. *The Anthropology of Music.* Evanston, IL: Northwestern University Press, 1964.
Milan, F. A., ed. *The Biology of Circumpolar Populations.* Cambridge: Cambridge University Press, 1980.
Miller, D., and D. R. Bjornson. "An Investigation of Cold Injured Soldiers in Alaska." *Military Medicine* 127 (1962): 247–252.

Mochanov, Yu. A. *Arkheologicheskie Pamyatniki Yakutii: Basseiny Aldana i Olekmy.* Novosibirsk: Nauka, 1983.

———. "The Bel'kachinsk Neolithic Culture on the Aldan." *Arctic Anthropology* 6, no. 1 (1969): 104–114.

———. *Drevneishie Etapy Zaseleniya Chelovekom Severo-Vostochoi Azii.* Novosibirsk: Nauka, 1977.

———. "The Ymyiakhtakh Late Neolithic Culture." *Arctic Anthropology* 6, no. 1 (1969): 115–118.

Mochanov, Yuri A., and Svetlana A. Fedoseeva. "Dyuktai Cave." In *American Beginnings: The Prehistory and Palaeoecology of Beringia*, edited by F. H. West, 164–174. Chicago: University of Chicago Press, 1996.

Mochanov, Yu. A., S. A. Fedoseeva, I. V. Konstantinov, N. V. Antipina, and A. G. Argunov. *Arkheologicheskie Pamyatniki Yakutii: Basseiny Vilyuya, Anabara i Oleneka.* Moscow: Nauka, 1991.

Monahan, C. M. "New Zooarchaeological Data from Bed II, Olduvai Gorge, Tanzania: Implications for Hominid Behavior in the Early Pleistocene." *Journal of Human Evolution* 31 (1996): 93–128.

Montet-White, Anta. *Le Malpas Rockshelter.* University of Kansas Publications in Anthropology no. 4. 1973.

Morlan, Richard E., and Jacques Cinq-Mars. "Ancient Beringians: Human Occupation in the Late Pleistocene of Alaska and the Yukon Territory." In *Paleoecology of Beringia*, edited by D. M. Hopkins, J. V. Matthews, Jr., C. E. Schweger, and S. B. Young, 353–381. New York: Academic Press, 1982.

Movius, Hallam L. "The Lower Paleolithic Cultures of Southern and Eastern Asia." *Transactions of the American Philosophical Society* 38 (1948): 329–420.

———. "A Wooden Spear of Third Interglacial Age from Lower Saxony." *Southwestern Journal of Anthropology* 6 (1950): 139–142.

Mulvaney, John, and Johan Kamminga. *Prehistory of Australia.* Washington, DC: Smithsonian Institution Press, 1999.

Murray, Maribeth S. "Local Heroes: The Long-Term Effects of Short-Term Prosperity—an Example from the Canadian Arctic." *World Archaeology* 30, no. 3 (1999): 466–483.

Napier, John. "The Antiquity of Human Walking." *Scientific American* 216, no. 4 (1967): 56–66.

Nettl, Bruno. *Music in Primitive Culture.* Cambridge: Harvard University Press, 1956.

Nielsen, J. M. *Armed Forces on a Northern Frontier: The Military in Alaska's History.* New York: Greenwood Press, 1988.

Nygaard, Signe E. "The Stone Age of Northern Scandinavia: A Review" *Journal of World Prehistory* 3, no. 1 (1989): 71–116.

Odess, Daniel, Stephen Loring, and William W. Fitzhugh. "Skraeling: First Peoples of Helluland, Markland, and Vinland." In *Vikings: The North Atlantic Saga*, edited by W. W. Fitzhugh and E. I. Ward, 193–205. Washington, DC: Smithsonian Institution Press, 2000.

Odner, Knut. *The Varanger Saami: Habitation and Economy AD 1200–1900*. Oslo: Scandinavian University Press, 1992.

Odum, Eugene P. *Ecology and Our Endangered Life-Support Systems*. Sunderland, MA: Sinauer Associates, 1993.

Ogilvie, Astrid E. J. "Documentary Evidence for Changes in the Climate of Iceland, A.D. 1500 to 1800." In *Climate since A.D. 1500,* edited by R. S. Bradley and P. D. Jones, 92–117. London: Routledge, 1992.

Okladnikov, A. P., and N. A. Beregovaya. *Drevnie Poseleniya Baranova Mysa*. Novosibirsk: Nauka, 1971.

Olsen, Haakon. "Osteologisk Materiale, Innledning: Fisk-Fugl. Varangerfunnene VI." *Tromsø Museums Skrifter* 7, no. 6 (1967).

Orr, K. D., and D. C. Fainer. 1952. "Cold Injuries in Korea during Winter of 1950–51." *Military Medicine* 31 (1952): 177–220.

Osgood, Cornelius. *Ingalik Material Culture*. Yale University Publications in Anthropology no. 22. 1940.

Oshibkina, S. V. "The Material Culture of the Veretye-type Sites in the Region to the East of Lake Onega." In *The Mesolithic in Europe*, edited by C. Bonsall, 402–413. Edinburgh: John Donald Publishers, 1989.

———. "Mezolit Tsentral'nykh i Severo-Vostochnykh Raionov Severa Evropeiskoi Chasti SSSR." In *Mezolit SSSR,* edited by L. V. Kol'tsov, 32–45. Moscow: Nauka, 1989.

Oswalt, Wendell H. *An Anthropological Analysis of Food-Getting Technology*. New York: John Wiley and Sons, 1976.

———. *Eskimos and Explorers*. 2nd ed. Lincoln: University of Nebraska Press, 1999.

———. "Technological Complexity: The Polar Eskimos and the Tareumiut." *Arctic Anthropology* 24, no. 2 (1987): 82–98.

Ovchinnikov, I. V., Götherström, A., Romanova, G. P., Kharitonov, V. M., Lidén, K., and Goodwin, W. "Molecular Analysis of Neanderthal DNA from the Northern Caucasus." *Nature* 404 (2000): 490–493.

Parés, J. M., and A. Pérez-González. "Magnetochronology and Stratigraphy at Gran Dolina Section, Atapuerca (Burgos, Spain)." *Journal of Human Evolution* 37 (1999): 325–342.

Parfitt, S. A., and M. B. Roberts. "Human Modification of Faunal Remains." In *Boxgrove: A Middle Pleistocene Hominid Site at Eartham Quarry, Boxgrove, West Sussex,* edited by M. B. Roberts and S. A. Parfitt, 395–415. English Heritage Archaeological Report no. 17. 1999.

Pavlov, Pavel, John Inge Svendsen, and Svein Indrelid. "Human Presence in the European Arctic nearly 40,000 Years Ago." *Nature* 413 (2001): 64–67.

Pennisi, Elizabeth. "Jumbled DNA Separates Chimps and Humans." *Science* 298 (2002): 719–721.

Péwé, Troy L. *Quaternary Geology of Alaska*. Geological Survey Professional Paper 835. 1975.

Pianka, Eric R. *Evolutionary Ecology*. 2nd ed. New York: Harper and Row Publishers, 1978.

Pidoplichko, I. G. *Mezhirichskie Zhilishcha iz Kostei Mamonta.* Kiev: Naukova Dumka, 1976.

———. *Pozdnepaleoliticheskie Zhilishcha iz Kostei Mamonta na Ukraine.* Kiev: Naukova Dumka, 1969.

Pilbeam, David. *The Ascent of Man: An Introduction to Human Evolution.* New York: Macmillan Publishing Co. 1969.

Pitul'ko, Vladimir V. "An Early Holocene Site in the Siberian High Arctic." *Arctic Anthropology* 30, no. 1 (1993): 13–21.

———. "Terminal Pleistocene—Early Holocene Occupation in Northeast Asia and the Zhokhov Assemblage." *Quaternary Science Reviews* 20, nos. 1–3 (2001): 267–275.

Pitul'ko, V. V., and A. K. Kasparov. "Ancient Arctic Hunters: Material Culture and Survival Strategy." *Arctic Anthropology* 33 (1996): 1–36.

Pitulko, V. V., P. A. Nikolski, E. Yu. Girya, A. E. Basilyan, V. E. Tumskoy, S. A. Koulakov, S. N. Astakhov, E. Yu. Pavlova, and M. A. Anisimov. "The Yana RHS Site: Humans in the Arctic before the Last Glacial Maximum." *Science* 303 (2004): 52–56.

Potts, R., and P. Shipman. "Cutmarks Made by Stone Tools on Bones from Olduvai Gorge, Tanzania." *Nature* 291 (1981): 577–580.

Powers, William Roger. "Paleolithic Man in Northeast Asia." *Arctic Anthropology* 10, no. 2 (1973): 1–106.

Powers, William R., and John F. Hoffecker. "Late Pleistocene Settlement in the Nenana Valley, Central Alaska." *American Antiquity* 54, no. 2 (1989): 263–287.

Powers, W. R., and R. H. Jordan. "Human Biogeography and Climate Change in Siberia and Arctic North America in the Fourth and Fifth Millennia BP." *Philosophical Transactions of the Royal Society London A* 330 (1990): 665–670.

Praslov, N. D. "Paleolithic Cultures in the Late Pleistocene." In *Late Quaternary Environments of the Soviet Union,* edited by A. A. Velichko, 313–318. Minneapolis: University of Minnesota Press, 1984.

Prokof'yeva, E. D. "The Nentsy." In *The Peoples of Siberia,* edited by M. G. Levin and L. P. Potapov, 547–570. Chicago: University of Chicago Press, 1964.

Puech, P. F., A. Prone, and R. Kraatz. "Microscopie de l'Usure Dentaire chez l'Homme Fossile: Bol Alimentaire et Environnement." *CRASP* 290 (1980): 1413–16.

Rainey, Froelich G. "Eskimo Prehistory: The Okvik Site on the Punuk Islands." *Anthropological Papers of the American Museum of Natural History* 37, no. 4 (1941): 443–569.

Raynal, Jean-Paul, Lionel Magoga, and Peter Bindon. "Tephrofacts and the First Human Occupation of the French Massif Central." In *The Earliest Occupation of Europe,* edited by W. Roebroeks and T. van Kolfschoten, 129–146. Leiden: University of Leiden, 1995.

Renouf, M.A.P. "Northern Coastal Hunter-Fishers: An Archaeological Model." *World Archaeology* 16, no. 1 (1984): 18–27.

Richards, Michael P., Paul B. Pettitt, Mary C. Stiner, and Erik Trinkaus. "Stable Iso-

tope Evidence for Increasing Dietary Breadth in the European Mid-Upper Paleolithic." *Proceedings of the National Academy of Sciences* 98 (2001): 6528–32.
Richards, M. P., P. B. Pettitt, E. Trinkaus, F. H. Smith, M. Paunovic, and I. Karavanic. "Neanderthal Diet at Vindija and Neanderthal Predation: The Evidence from Stable Isotopes." *Proceedings of the National Academy of Sciences* 97, no. 13 (2000): 7663–66.
Rightmire, G. Philip. *The Evolution of* Homo erectus: *Comparative Anatomical Studies of an Extinct Human Species*. Cambridge: Cambridge University Press, 1990.
———. "Human Evolution in the Middle Pleistocene: The Role of *Homo heidelbergensis.*" *Evolutionary Anthropology* 6, no. 6 (1998): 218–227.
———. "Patterns of Hominid Evolution and Dispersal in the Middle Pleistocene." *Quaternary International* 75 (2001): 77–84.
Roberts, D. F. "Body Weight, Race, and Climate." *American Journal of Physical Anthropology* 11 (1953): 533–558.
———. *Climate and Human Variability*. Reading: Addison-Wesley, 1973.
Roberts, M. B., and S. A. Parfitt, eds. *Boxgrove: A Middle Pleistocene Hominid Site at Eartham Quarry, Boxgrove, West Sussex*. English Heritage Archaeological Report no. 17. 1999.
Roberts, M. B., C. B. Stringer, and S. A. Parfitt. "A Hominid Tibia from Middle Pleistocene Sediments at Boxgrove, UK." *Nature* 369 (1994): 311–313.
Roberts, Neil. *The Holocene: An Environmental History*. 2nd ed. Oxford: Blackwell Publishers, 1998.
Roberts, R. G., R. Jones, N. A. Spooner, M. J. Head, A. S. Murray, and M. A. Smith. "The Human Colonisation of Australia: Optical Dates of 53,000 and 60,000 Years Bracket Human Arrival at Deaf Adder Gorge, Northern Territory." *Quaternary Science Reviews* 13 (1994): 575–586.
Robinson, John T. "Adaptive Radiation in the Australopithecines and the Origin of Man." In *African Ecology and Human Evolution,* edited by F. C. Howell and F. Bourliere, 385–416. Chicago: Aldine, 1963.
Rodman, Peter S., and Henry M. McHenry. "Bioenergetics of Hominid Bipedalism." *American Journal of Physical Anthropology* 52 (1980): 103–106.
Roe, Derek. "The Orce Basin (Andalucia, Spain) and the Initial Palaeolithic of Europe." *Oxford Journal of Archaeology* 14 (1995): 1–12.
Roebroeks, W. "Archaeology and Middle Pleistocene Stratigraphy: The Case of Maastricht-Belvédère (NL)." In *Chronostratigraphie et Faciès Culturels du Paléolithique Inférieur et Moyen dans l'Europe de Nord-Ouest,* edited by A. Tuffreau and J. Somme, 81–86. Paris: Supplement au Bulletin de l'Association Française pour l'Étude du Quaternaire, 1986.
Roebroeks, Wil, Nicholas J. Conard, and Thijs van Kolfschoten. "Dense Forests, Cold Steppes, and the Palaeolithic Settlement of Northern Europe." *Current Anthropology* 33 (1992): 551–586.
Roebroeks, Wil, J. Kolen, and E. Rensink. "Planning Depth, Anticipation, and the Organization of Middle Palaeolithic Technology: The 'Archaic Natives' Meet Eve's Descendents." *Helinium* 28, no. 1 (1988): 17–34.

Roebroeks, Wil, and Thijs van Kolfschoten. "The Earliest Occupation of Europe: A Reappraisal of Artefactual and Chronological Evidence." In *The Earliest Occupation of Europe*, edited by W. Roebroeks and T. van Kolfschoten, 297–315. Leiden: University of Leiden, 1995.

Rogachev, A. N., and Sinitsyn, A. A. "Kostenki 15 (Gorodtsovskaya Stoyanka)." In *Paleolit Kostenkovsko-Borshchevskogo Raiona na Donu 1879–1979*, edited by N. D. Praslov and A. N. Rogachev, 162–171. Leningrad: Nauka, 1982.

Rolland, Nicholas, and Harold L. Dibble. "A New Synthesis of Middle Paleolithic Variability." *American Antiquity* 55 (1990): 480–499.

Roth, Henry Lee. *The Aborigines of Tasmania*. London: Kegan Paul, Trench, Trubner, 1890.

Rudenko, S. I. "The Ancient Culture of the Bering Sea and the Eskimo Problem." *Arctic Institute of North America, Anthropology of the North, Translations from Russian Sources*, no. 1. 1961.

Ruff, Christopher. "Climate, Body Size, and Body Shape in Human Evolution." *Journal of Human Evolution* 21 (1991): 81–105.

Sablin, Mikhail V., and Gennady A. Khlopachev. "The Earliest Ice Age Dogs: Evidence from Eliseevichi I." *Current Anthropology* 43, no. 5 (2002): 795–799.

Sahlins, Marshall. *Culture and Practical Reason*. Chicago: University of Chicago Press, 1976.

Sahnouni, M., and J. de Heinzelin. "The Site of Aïn Hanech Revisited: New Investigations at This Lower Pleistocene Site in Northern Algeria." *Journal of Archaeological Science* 25 (1998): 1083–1101.

Saragusti, Idit, and Naama Goren-Inbar. "The Biface Assemblage from Gesher Benot Ya'aqov, Israel: Illuminating Patterns in 'Out of Africa' Dispersal." *Quaternary International* 75 (2001): 85–89.

Sarich, V. M., and A. C. Wilson. "Immunological Time Scale for Hominid Evolution." *Science* 158 (1967): 1200–1203.

Schaffel, Kenneth. *The Emerging Shield: The Air Force and the Evolution of Continental Air Defense, 1945–1960*. Washington, DC: Office of Air Force History, 1991.

Schick, Kathy D., and Nicholas Toth. *Making Silent Stones Speak: Human Evolution and the Dawn of Technology*. New York: Simon and Schuster, 1993.

Schledermann, Peter. "Ellesmere: Vikings in the Far North." In *Vikings: The North Atlantic Saga*, edited by W. W. Fitzhugh and E. I. Ward, 248–256. Washington, DC: Smithsonian Institution Press, 2000.

———. "Preliminary Results of Archaeological Investigations in the Bache Peninsula Region, Ellesmere Island, N.W.T." *Arctic* 31, no. 4 (1978): 459–474.

Scott, G. Richard, Scott Legge, Robert W. Lane, Susan L. Steen, and Steven R. Street. "Physical Anthropology of the Arctic." In *The Arctic: Environment, People, Policy*, edited by M. Nuttall and T. V. Callaghan, 339–373. Amsterdam: Harwood Academic Publishers, 2000.

Scott, Katherine. "Mammoth Bones Modified by Humans: Evidence from La Cotte de St. Brelade, Jersey, Channel Islands." In *Bone Modification*, edited by R. Bonnichsen and M. H. Sorg, 335–346. Orono, ME: Center for the Study of the First Americans, 1989.

Semenov, S. A. *Prehistoric Technology*. Translated by M. W. Thompson. New York: Barnes and Noble, 1964.

Shackleton, N. J., and N. D. Opdyke. "Oxygen Isotope and Paleomagnetic Stratigraphy of Equatorial Pacific Core V28-238: Temperatures and Ice Volumes on a 10^3 and 10^6 Year Scale." *Quaternary Research* 3 (1973): 39–55.

Shea, John J. "Spear Points from the Middle Paleolithic of the Levant." *Journal of Field Archaeology* 15 (1988): 441–450.

Shen, Guanjun, Wei Wang, Qian Wang, Jianxin Zhao, Kenneth Collerson, Chunlin Zhou, and Phillip V. Tobias. "U-Series Dating of Liujiang Hominid Site in Guangxi, Southern China." *Journal of Human Evolution* 43 (2002): 817–829.

Shipman, Pat. "Scavenging or Hunting in Early Hominids." *American Anthropologist* 88 (1986): 27–43.

Shipman, P., and J. Rose. "Evidence of Butchery and Hominid Activities at Torralba and Ambrona: An Evaluation Using Microscopic Techniques." *Journal of Archaeological Science* 10 (1983): 465–474.

Shovkoplyas, I. G. *Mezinskaya Stoyanka*. Kiev: Naukova Dumka, 1965.

Simonsen, Povl. "Varanger-Funnene II. Fund og Udgravninger pa Fjordens Sydkyst." *Tromsø Museums Skrifter* 7, no. 2 (1961).

———. "When and Why Did Occupational Specialization Begin at the Scandinavian North Coast?" In *Prehistoric Maritime Adaptations of the Circumpolar Zone*, edited by W. Fitzhugh, 75–85. The Hague: Mouton Publishers, 1975.

Singer, Ronald, Bruce G. Gladfelter, and John J. Wymer. *The Lower Paleolithic Site at Hoxne, England*. Chicago: University of Chicago Press, 1993.

Sinitsyn, A. A. "Nizhnie Kul'turnye Sloi Kostenok 14 (Markina Gora) (Raskopki 1998–2001 gg.)." In *Kostenki v Kontekste Paleolita Evrazii*, edited by A. A. Sinitsyn, V. Ya. Sergin, and J. F. Hoffecker, 219–236. Saint Petersburg: Russian Academy of Sciences, 2002.

Soffer, Olga. *The Upper Paleolithic of the Central Russian Plain*. San Diego: Academic Press, 1985.

Soffer, Olga, J. M. Adovasio, and D. C. Hyland. "The 'Venus' Figurines: Textiles, Basketry, Gender, and Status in the Upper Paleolithic." *Current Anthropology* 41, no. 4 (2000): 511–537.

Soffer, Olga, Pamela Vandiver, Bohuslav Klima, and Jiří Svoboda. "The Pyrotechnology of Performance Art: Moravian Venuses and Wolverines." In *Before Lascaux: The Complex Record of the Early Upper Paleolithic*, edited by H. Knecht, A. Pike-Tay, and R. White, 259–275. Boca Raton, FL: CRC Press, 1993.

Solbakk, Aage, ed. *The Sami People*. Karasjok, Norway: Sámi Instituhtta/Davvi Girji O.S., 1990.

Solecki, Ralph S. *Shanidar, the First Flower People*. New York: Alfred Knopf, 1971.

Sommer, Jeffrey D. "The Shanidar IV 'Flower Burial': A Re-evaluation of Neanderthal Burial Ritual." *Cambridge Archaeological Journal* 9, no. 1 (1999): 127–129.

Spoonheimer, Matt, and Julia A. Lee-Thorp. "Isotopic Evidence for the Diet of an Early Hominid, *Australopithecus africanus*." *Science* 283 (1999): 368–370.

Spoor, F., B. Wood, and F. Zonneveld. "Implications of Early Hominid Labyrinthine Morphology for Evolution of Human Bipedal Locomotion." *Nature* 369 (1994): 645–648.

Stanford, Dennis J. *The Walakpa Site, Alaska: Its Place in the Birnirk and Thule Cultures.* Smithsonian Contributions to Anthropology no. 20. 1976.

Stiner, Mary C. *Honor among Thieves: A Zooarchaeological Study of Neandertal Ecology.* Princeton: Princeton University Press, 1994.

Stiner, M. C., N. D. Munro, T. A. Surovell, E. Tchernov, and O. Bar-Yosef. "Paleolithic Population Growth Pulses Evidenced by Small Animal Exploitation." *Science* 283 (1999): 190–194.

Straus, Lawrence Guy. *Iberia before the Iberians: The Stone Age Prehistory of Cantabrian Spain.* Albuquerque: University of New Mexico Press, 1992.

———. "The Upper Paleolithic of Europe: An Overview." *Evolutionary Anthropology* 4, no. 1 (1995): 4–16.

Stringer, C. B. "Secrets of the Pit of the Bones." *Nature* 362 (1993): 501–502.

Stringer, C. B., and E. Trinkaus. "The Human Tibia from Boxgrove." In *Boxgrove: A Middle Pleistocene Hominid Site at Eartham Quarry, Boxgrove, West Sussex,* edited by M. B. Roberts and S. A. Parfitt, 420–422. English Heritage Archaeological Report no. 17. 1999.

Stringer, Christopher, and Clive Gamble. *In Search of the Neanderthals: Solving the Puzzle of Human Origins.* New York: Thames and Hudson, 1993.

Stringer, Christopher, and Robin McKie. *African Exodus: The Origins of Modern Humanity.* New York: Henry Holt and Co., 1996.

Sugiyama, Yukimaru. "Social Tradition and the Use of Tool-Composites by Wild Chimpanzees." *Evolutionary Anthropology* 6, no. 1 (1997): 23–27.

Sulimirski, Tadeusz. *Prehistoric Russia: An Outline.* London: John Baker, 1970.

Sumner, D. S., T. L. Criblez, and W. H. Doolittle. "Host Factors in Human Frostbite." *Military Medicine* 139 (1974): 454–461.

Susman, R. L., and J. T. Stern. "Functional Morphology of *Homo habilis.*" *Science* 217 (1982): 931–934.

Susman, R. L., J. T. Stern, and W. L. Jungers. "Arboreality and Bipedality in the Hadar Hominids." *Folia Primatologica* 43 (1984): 113–156.

Sutherland, Patricia. 2000. "The Norse and Native North Americans." In *Vikings: The North Atlantic Saga,* edited by W. W. Fitzhugh and E. I. Ward, 238–247. Washington, DC: Smithsonian Institution Press, 2000.

Svoboda, Jiří, Vojen Ložek, and Emanuel Vlček. *Hunters between East and West: The Paleolithic of Moravia.* New York: Plenum Press, 1996.

Swisher, C. C., G. H. Curtis, T. Jacob, A. G. Getty, A. Suprijo, and Widiasmoro. "Age of the Earliest Known Hominids in Java, Indonesia." *Science* 263 (1994): 1118–21.

Swisher, Carl C., Garniss H. Curtis, and Roger Lewin. *Java Man.* New York: Scribner, 2000.

Szathmary, Emöke J. E. "Human Biology of the Arctic." In *Handbook of the North American Indian,* vol. 5, *Arctic,* edited by D. Damas, 64–71. Washington, DC: Smithsonian Institution, 1984.

Tarasov, L. M. *Gagarinskaya Stoyanka i Ee Mesto v Paleolite Evropy.* Leningrad: Nauka, 1979.
Tattersall, Ian. *The Fossil Trail: How We Know What We Think We Know about Human Evolution.* New York: Oxford University Press, 1995.
———. *The Last Neanderthal: The Rise, Success, and Mysterious Extinction of Our Closest Human Relatives.* Rev. ed. Boulder, CO: Westview Press, 1999.
Taylor, R. E. "Radiocarbon Dating: The Continuing Revolution." *Evolutionary Anthropology* 4 (1996): 169–181.
Taylor, William E., and George Swinton. "Prehistoric Dorset Art." *The Beaver* (winter 1967): 32–47.
Teleki, Geza. *The Predatory Behavior of Wild Chimpanzees.* Lewisburg, PA: Bucknell University Press, 1973.
Thieme, Hartmut. "Altpaläolithische Wurfspeere aus Schöningen, Niedersachsen: Ein Vorbericht." *Archäologisches Korrespondenzblatt* 26 (1996): 377–393.
———. "Lower Palaeolithic Hunting Spears from Germany." *Nature* 385 (1997): 807–810.
Thomson, A., and L.H.D. Buxton. "Man's Nasal Index in Relation to Certain Climatic Conditions." *Journal of the Royal Anthropological Institute* 53 (1923): 92–122.
Titon, Jeff Todd, James T. Koetting, David P. McAllester, David B. Reck, and Mark Slobin. *Worlds of Music: An Introduction to the Music of the World's Peoples.* New York: Schirmer Books, 1984.
Torrence, Robin. "Time Budgeting and Hunter-Gatherer Technology." In *Hunter-Gatherer Economy in Prehistory: A European Perspective*, edited by G. Bailey, 11–22. Cambridge: Cambridge University Press, 1983.
Toth, Nicholas. "The Oldowan Reassessed: A Close Look at Early Stone Artifacts." *Journal of Archaeological Science* 12 (1985): 101–120.
Trigger, Bruce G. *A History of Archaeological Thought.* Cambridge: Cambridge University Press, 1989.
Trinkaus, Erik. "Bodies, Brawn, Brains, and Noses: Human Ancestors and Human Predation." In *The Evolution of Human Hunting*, edited by M. Nitecki and D. V. Nitecki, 107–145. New York: Plenum Press, 1987.
———. "Neanderthal Limb Proportions and Cold Adaptation." In *Aspects of Human Evolution*, edited by C. Stringer, 187–224. London: Taylor and Francis, 1981.
———. *The Shanidar Neandertals.* New York: Academic Press, 1983.
Trinkaus, Erik, et al. "An Early Modern Human from the Pestera cu Oase, Romania." *Proceedings of the National Academy of Sciences* 100, no. 20 (2003): 11231–36.
Trinkaus, Erik, and Pat Shipman. *The Neandertals: Of Skeletons, Scientists, and Scandal.* New York: Vintage Books, 1994.
Troeng, John. *Worldwide Chronology of Fifty-three Prehistoric Innovations.* Acta Archaeologica Lundensia 8, no. 21 (1993).
Tuffreau, Alain, and Pierre Antoine. "The Earliest Occupation of Europe: Continental Northwestern Europe." In *The Earliest Occupation of Europe*, edited by

W. Roebroeks and T. van Kolfschoten, 147–163. Leiden: University of Leiden, 1995.

Tuffreau, A., P. Antoine, P. Chase, H. L. Dibble, B. B. Ellwood, Th. Van Kolfschoten, A. Lamotte, M. Laurent, S. P. McPherron, A.-M. Moigne, and A. V. Munaut. "Le Gisement Acheuléen de Cagny-L'Epinette (Somme)." *Bulletin de la Société Préhistorique Française* 92 (1995): 169–199.

Turner, Alan. "Large Carnivores and Earliest European Hominids: Changing Determinants of Resource Availability during the Lower and Middle Pleistocene." *Journal of Human Evolution* 22 (1992): 109–126.

Turner, Christy G. "Teeth, Needles, Dogs, and Siberia: Bioarchaeological Evidence for the Colonization of the New World." In *The First Americans*, 123–158. Memoirs of the California Academy of Sciences no. 27. 2002.

Turner, Elaine. "The Problems of Interpreting Hominid Subsistence Strategies at Lower Palaeolithic Sites—a Case Study from the Central Rhineland of Germany." In *Hominid Evolution: Lifestyles and Survival Strategies*, edited by H. Ullrich, 365–382. Gelsenkirchen-Schwelm: Edition Archaea, 1999.

Turner, Victor. *The Forest of Symbols: Aspects of Ndembu Ritual*. Ithaca, NY: Cornell University Press, 1967.

Tuttle, Russell H. *Apes of the World: Their Social Behavior, Communication, Mentality, and Ecology*. Park Ridge, NJ: Noyes Publications, 1986.

———. "Knuckle-walking and the Problems of Human Origins." *Science* 166 (1969): 953–961.

Valoch, Karel. "The Earliest Occupation of Europe: Eastern Central and Southeastern Europe." In *The Earliest Occupation of Europe*, edited by W. Roebroeks and T. van Kolfschoten, 67–84. Leiden: University of Leiden, 1995.

Vandiver, Pamela P., Olga Soffer, Bohuslav Klima, and Jiři Svoboda. "The Origins of Ceramic Technology at Dolni Vestonice, Czechoslovakia." *Science* 246 (1989): 1002–1008.

Vasil'ev, Sergey A. "The Final Paleolithic in Northern Asia: Lithic Assemblage Diversity and Explanatory Models." *Arctic Anthropology* 38, no. 2 (2001): 3–30.

Vekua, Abesalom, et al. "A New Skull of Early *Homo* from Dmanisi, Georgia." *Science* 297 (2002): 85–89.

Velichko, A. A. "Late Pleistocene Spatial Paleoclimatic Reconstructions." In *Late Quaternary Environments of the Soviet Union*, edited by A. A. Velichko, 261–285. Minneapolis: University of Minnesota Press, 1984.

———. "Loess-Paleosol Formation on the Russian Plain." *Quaternary International* 7/8 (1990): 103–114.

Velichko, A. A., A. B. Bogucki, T. D. Morozova, V. P. Udartsev, T. A. Khalcheva, and A. I. Tsatkin. "Periglacial Landscapes of the East European Plain." In *Late Quaternary Environments of the Soviet Union*, edited by A. A. Velichko, 94–118. Minneapolis: University of Minnesota Press, 1984.

Vereshchagin, N. K., and G. F. Baryshnikov. "Paleoecology of the Mammoth Fauna in the Eurasian Arctic." In *Paleoecology of Beringia*, edited by D. M. Hopkins, J. V. Matthews, C. E. Schweger, and S. B. Young, 267–279. New York: Academic Press, 1982.

Vereshchagin, N. K., and Kuz'mina, I. E. "Ostatki Mlekopitayushchikh iz Paleoliticheskikh Stoyanok na Donu i Verkhnei Desne." *Trudy Zoologicheskogo Instituta AN SSSR* 72 (1977): 77–110.

Villa, Paola. "Early Italy and the Colonization of Western Europe." *Quaternary International* 75 (2001): 113–130.

———. *Terra Amata and the Middle Pleistocene Archaeological Record of Southern France.* Berkeley: University of California Press, 1983.

Villa, P., and F. Bon. "Fire and Fireplaces in the Lower, Middle and Early Upper Paleolithic of Western Europe." *Journal of Human Evolution* 42, no. 3 (2002): A37–A38.

Vlček, Emanuel. "Patterns of Human Evolution." In *Hunters between East and West: The Paleolithic of Moravia*, by J. Svoboda, V. Ložek, and E. Vlček, 37–74. New York: Plenum Press, 1996.

Vorren, Ørnulv. *Norway North of 65*. Oslo: Oslo University Press, 1960.

Wahlgren, Erik. *The Vikings and America*. London: Thames and Hudson, 1986.

Walker, Alan, and Pat Shipman. *The Wisdom of the Bones: In Search of Human Origins.* New York: Alfred A. Knopf, 1996.

Walker, Alan, and Mark Teaford. "Inferences from Quantitative Analysis of Dental Microwear." *Folia Primatologica* 53 (1989): 177–189.

Waters, Michael R., Steven L. Forman, and James M. Pierson. "Diring Yuriakh: A Lower Paleolithic Site in Central Siberia." *Science* 275 (1997): 1281–1284.

West, Frederick Hadleigh. *The Archaeology of Beringia*. New York: Columbia University Press, 1981.

West, Frederick Hadleigh, ed. *American Beginnings: The Prehistory and Palaeoecology of Beringia*. Chicago: University of Chicago Press, 1996.

Whallon, Robert. "Elements of Cultural Change in the Later Paleolithic." In *The Human Revolution*, edited by P. Mellars and C. Stringer, 433–454. Princeton: Princeton University Press, 1989.

White, Leslie A. "On the Use of Tools by Primates." *Journal of Comparative Psychology* 34 (1942): 369–374.

———. *The Science of Culture: A Study of Man and Civilization*. New York: Grove Press, 1949.

Whittaker, Robert H. *Communities and Ecosystems*. 2nd ed. New York: Macmillan Publishing Co., 1975.

Wiessner, Polly. "Risk, Reciprocity, and Social Influences on !Kung San Economics." In *Politics and History in Band Societies*, edited by E. Leacock and R. B. Lee, 61–84. Cambridge: Cambridge University Press, 1982.

Wilmsen, Edwin N. "Interaction, Spacing Behavior, and the Organization of Hunting Bands." *Journal of Anthropological Research* 29 (1973): 1–31.

Winterhalder, Bruce. "Diet Choice, Risk, and Food Sharing in a Stochastic Environment." *Journal of Anthropological Archaeology* 5 (1986): 369–392.

Woillard, G. "Grande Pile Peat Bog: A Continuous Pollen Record for the Last 140,000 Years." *Quaternary Research* 9 (1978): 1–21.

Wolpoff, Milford H. *Paleoanthropology*. 2nd ed. Boston: McGraw-Hill, 1999.

Wood, B. "Origin and Evolution of the Genus *Homo*." *Nature* 355 (1982): 783–790.

Wright, Jr., H. E., J. E. Kutzbach, T. Webb III, W. F. Ruddimann, F. A. Street-Perrott, and P. J. Bartlein, eds. *Global Climates since the Last Glacial Maximum.* Minneapolis: University of Minnesota Press, 1993.

Wymer, John J., and Ronald Singer. "Flint Industries and Human Activity." In *The Lower Paleolithic Site at Hoxne, England,* edited by R. Singer, B. G. Gladfelter, and J. J. Wymer, 74–128. Chicago: University of Chicago Press, 1993.

Wynn, Thomas G. "The Evolution of Tools and Symbolic Behaviour." In *Handbook of Human Symbolic Evolution,* edited by Andrew Lock and Charles R. Peters, 263–287. Oxford: Blackwell Publishers, 1999.

———. "Handaxe Enigmas." *World Archaeology* 27, no. 1 (1995): 10–24.

———. "Piaget, Stone Tools, and the Evolution of Human Intelligence." *World Archaeology* 17, no. 1 (1985): 32–43.

Yesner, David R. "Human Dispersal into Interior Alaska: Antecedent Conditions, Mode of Colonization, and Adaptations." *Quaternary Science Reviews* 20, nos. 1–3 (2001): 315–327.

Young, Steven B. *To the Arctic: An Introduction to the Far Northern World.* New York: John Wiley and Sons, 1994.

Zagorska, Ilga, and Francis Zagorskis. "The Bone and Antler Inventory from Zvejnieki II, Latvian SSR." In *The Mesolithic in Europe,* edited by C. Bonsall, 414–423. Edinburgh: John Donald Publishers, 1989.

Zaloga, S. J. *Target America: The Soviet Union and the Strategic Arms Race, 1945–1964.* Novato: Presidio Press, 1993.

Zavernyaev, F. M. *Khotylevskoe Paleoliticheskoe Mestonakhozhdenie.* Leningrad: Nauka, 1978.

Zhu, R. X., K. A. Hoffman, R. Potts, C. L. Deng, Y. X. Pan, B. Guo, C. D. Shi, Z. T. Guo, B. Y. Yuan, Y. M. Hou, and W. W. Huang. "Earliest Presence of Humans in Northeast Asia." *Nature* 413 (2001): 413–417.

Figure Credits

2.4. Redrawn from Alan Walker and Pat Shipman, *The Wisdom of the Bones: In Search of Human Origins* (New York: Alfred A. Knopf, 1996), p. 245.

2.5. Redrawn from Mary D. Leakey, *Olduvai Gorge: Excavations in Beds I and II, 1960–1963* (Cambridge: Cambridge University Press, 1971), figs. 10.4, 14.3, 16.1–16.3.

2.7. Redrawn from Alan Walker and Pat Shipman, *The Wisdom of the Bones: In Search of Human Origins* (New York: Alfred A. Knopf, 1996), p. 245.

2.8. Redrawn from Mary D. Leakey, *Olduvai Gorge: Excavations in Beds I and II, 1960–1963* (Cambridge: Cambridge University Press, 1971), fig. 97.

3.1. Based on Ann Henderson-Sellers and Peter J. Robinson, *Contemporary Climatology* (Edinburgh Gate: Addison Wesley Longman, 1986), fig. 2.32.

3.4. Drawn from a photograph in C. B. Stringer and E. Trinkaus, "The Human Tibia from Boxgrove," in *Boxgrove: A Middle Pleistocene Hominid Site at Eartham Quarry, Boxgrove, West Sussex*, ed. M. B. Roberts and S. A. Parfit, pp. 420–422 (English Heritage Archaeological Report no. 17, 1999), fig. 339.

3.5. Redrawn from Hartmut Thieme, "Altpaläolithische Wurfspeere aus Schöningen, Niedersachsen: Ein Vorbericht," *Archäologisches Korrespondenzblatt* 26 (1996): fig. 9.

4.1. Based on G. Philip Rightmire, "Human Evolution in the Middle Pleistocene: The Role of *Homo heidelbergensis*," *Evolutionary Anthropology* 6, no. 6 (1998): fig. 2.

4.2. Redrawn from Steven Emilio Churchill, "Cold Adaptation, Heterochrony, and Neandertals," *Evolutionary Anthropology* 7, no. 2 (1998): fig. 1.

4.4. Prepared core and flake redrawn from François Bordes, *The Old Stone Age*, trans. J. E. Anderson (New York: McGraw-Hill Book Co., 1998), fig. 8; reconstruction of composite flake tool based on Patricia Anderson-Gerfaud, "Aspects of Behaviour in the Middle Palaeolithic: Functional Analysis of Stone Tools from Southwest France," in *The Emergence of Modern Humans*, ed. P. Mellars, pp. 389–418 (Edinburgh: Edinburgh University Press, 1990), fig. 14.

4.5. Drawn from a photograph in G. Baryshnikov and J. F. Hoffecker, "Mousterian Hunters of the NW Causacus: Preliminary Results of Recent Investigations," *Journal of Field Archaeology* 21 (1994): fig. 5.

4.6. Photograph by the author.

5.1. Redrawn from P. Lieberman, J. T. Laitman, J. S. Reidenberg, and P. J. Gannon, "The Anatomy, Physiology, Acoustics, and Perception of Speech: Essential Elements in the Analysis of the Evolution of Human Speech," *Journal of Human Evolution* 23 (1998): 447–467.

5.3. Photograph by the author.

5.4. Based on data in Erik Trinkaus, "Neanderthal Limb Proportions and Cold

Adaptation," in *Aspects of Human Evolution*, ed. C. Stringer (London: Taylor and Francis, 1981), table 7.

5.5. Redrawn from A. A. Sinitsyn, "Nizhnie Kul'turnye Sloi Kostenok 14 (Markina Gora) (Raskopki 1998–2000 gg.)," in *Kostenki v. Kontekste Paleolita Evrazii*, ed. A. A. Sinitsyn, V. Ya. Sergin, and J. F. Hoffecker (Saint Petersburg: Russian Academy of Sciences, 2002), fig. 9.

5.6. Redrawn from O. N. Bader, "Pogrebeniya v Verkhnem Paleolite i Mogila na Stoyanke Sungir'," *Sovetskaya Arkheologiya* 3 (1967): 142–159.

5.7. Redrawn from Chester S. Chard, *Northeast Asia in Prehistory* (Madison: University of Wisconsin, 1974), fig. 1.13.

5.8. Redrawn from P. P. Efimenko, *Kostenki I* (Moscow: USSR Academy of Sciences, 1958), fig. 11.

5.9. Photograph by the author.

6.1. Redrawn from Paul G. Bahn and Jean Vertut, *Journey through the Ice Age* (Berkeley: University of California Press, 1997), figs. 7.16–7.17.

6.3. Modified from I. G. Pidoplichko, *Pozdnepaleoliticheskie Zhilishcha iz Kostei Mamonta na Ukraine* (Kiev: Naukova Dumka, 1969), fig. 43.

6.4. Redrawn from Grigoriy M. Burov, "Some Mesolithic Wooden Artifacts from the Site of Vis I in the European North East of the U.S.S.R.," in *The Mesolithic in Europe*, ed. C. Bonsall (Edinburgh: John Donald Publishers, 1989), fig. 6.

6.5. Redrawn from Yu. A. Mochanov, *Arkheologicheskie Pamyatniki Yakutii: Basseiny Aldana i Olëkmy* (Novosibirsk: Nauka, 1983), fig. 169.

6.7. Photograph by Charles E. Holmes.

6.8. Redrawn from N. N. Dikov, *Arkheologicheskie Pamyatniki Kamchatki, Chukotki i Verkhnei Kolymy* (Moscow: Nauka, 1977), fig. 12.

7.2. Redrawn from Roberto Bosi, *The Lapps* (London: Thames and Hudson, 1960), fig. 10.

7.3. Redrawn from Roberto Bosi, *The Lapps* (London: Thames and Hudson, 1960), fig. 34.

7.5. Redrawn from Yu. A. Mochanov, *Arkheologicheskie Pamyatniki Yakutii: Basseiny Aldena i Olëkmy* (Novosibirsk: Nauka, 1983), fig. 87.

7.7. Redrawn from photograph in Robert McGhee, *Canadian Arctic Prehistory* (Hull, Quebec: Canadian Museum of Civilization, 1990), plate 3.

7.9. Redrawn from Helge Larsen and Froelich Rainey, *Ipiutak and the Arctic Whale Hunting Culture*, Anthropological Papers of the American Museum of Natural History, vol. 42 (1948), fig. 39.

7.10. Photograph taken by A. M. Bailey, March 1922. All Rights Reserved. Image Archives, Denver Museum of Nature and Science.

7.11. Photograph by Mandy Whorton.

B1. Adapted from N. J. Shackleton and N. D. Opdyke, "Oxygen Isotope and Paleomagnetic Stratigraphy of Equatorial Pacific Core V28–238: Temperatures and Ice Volumes on a 10^3 and 10^6 Year Scale," *Quaternary Research* 3 (1973): 39–55.

B2. Adapted from N. J. Shackleton and N. D. Opdyke, "Oxygen Isotope and Paleomagnetic Stratigraphy of Equatorial Pacific Core V28–238: Temperatures and Ice Volumes on a 10^3 and 10^6 Year Scale," *Quaternary Research* 3 (1973): 39–55.

B3. Adapted from W. Dansgaard et al., "Evidence for General Instability of Past Climate from a 250-kyr Ice-Core Record," *Nature* 364 (1993): fig. 1.

B4. Adapted from T. C. Atkinson, K. R. Briffa, and G. R. Coope, "Seasonal Temperatures in Britain during the Last 22,000 Years, Reconstructed Using Beetle Remains," *Nature* 325 (1987): fig. 2c.

B5. Adapted from J. T. Andrews et al., "Relative Departures in July Temperatures in Northern Canada for the Past 6,000 Yr.," *Nature* 289 (1981): fig. 3.

Index

Note: Page numbers in *italics* denote illustrations.

Abri Blanchard, 84
Abric Romani, 60. *See also* wood technology
Acheulean industry, 23, 25, 41, 53, 60, 153n54
adzes, 101, 126–127
Afontova Gora, 111
agriculture and farming, 120, 125, 130
Aïn Hanech, 23
Alaska: Cold War in, 141–142; Denbigh Flint complex in, 127–128; Denali complex of interior, 117, 119, 176n82; Ice Age (Pleistocene) mammals in, 114–115; initial settlement of, 115–117; later prehistory of, 134–139. *See also* Beringia
albedo, 113
Aldan River, 112, 125
Aleutian Islands, 119, 121
Allen rule, 150n26. *See also* Bergmann rule; biogeographic "rules"
Alta Fjord, 123
Altai, 58, 77
Altamira, 82
amber, 105
Americas, peopling of, 8, 96, 112
Amud Cave, 156n36
anatomical adaptations to cold: *Homo heidelbergensis*, 36–38; Inuit, 5; modern human (Upper Paleolithic), 8, 96, 104, 111, 115; Neanderthal, 8, 47, 49, 53–55, *54*, 59, 70; postglacial Europeans, 109; Saami, 103
anatomical adaptations to heat: *Homo erectus*, 25, 72; modern humans, 8, 68–72, 95
anatomy of speech, 66, 72–73, *73*, 76. *See also* basicranium; language; modern humans; vocal tract
Andaman Islanders, 157n50
Antarctica, 6, 50

Arabian Peninsula, 18
Arago, 36, 46, 49, 152n40
Arctic: archaeological record of, 125; defined in terms of climate, 174n64; foraging peoples of, 70–71, 91–92; industrial settlement of, 9, 140–142; initial settlement of, 8, 85, 96–97, 101, 109; Vikings in, 1–4
Arctic Circle, 1, 9, 101, 113, 120, 132; settlement above, 81, 96, 109, 121
arctic hare. *See* hare
Arctic Small Tool tradition, 130, 132, 134
Ariendorf, 46
art: cave paintings, 82, 100, 162n41; Dorset culture, 133; Ipiutak culture, 138; Komsa culture, 103; Late Stone Age (Norway), 122–124, *123;* modern human (Upper Paleolithic), 71, 73, 76, 76, *83,* 87–89, *89,* 91, 100, 106, 162–163n41; Neanderthal, lack of, 66; northern Russia, 109; Old Bering Sea/Okvik cultures, 136; Siberian, 110; structural complexity of, 88, 165n72. *See also* "Venus" figurines
Artemisia, 23
Atapuerca, 31–32, 49–50, 149n13
Athapaskans, 107, 172n39
Atlantic period, 98, 121, 125, 128, 130, *131,* 133
aurochs, 107
Australia, 70, 75–76, 84, 86, 89, 165n74
australopithecines, 15–22, 24, 27, 146n29; *Ardipithecus ramidus,* 15; *Australopithecus afarensis, 16; Australopithecus africanus,* 19
Avdeevo, 92–93
awls, 62, 111

Baffin Island, 1, 140
Barakaevskaya Cave, 67
Barents Sea, 109, 125

215

basal metabolic rate, 63
basicranium, 66, 72, 76
bear, 32, 112, 133. *See also* polar bear
Beaufort Sea, 176n82
beetles, fossil, as indicator of past temperatures, 39–40, 78, 98, *99,* 130
Bel'kachi I, 125
Bel'kachinsk culture, 127, 129
Berelëkh, 175n72
Bergmann rule, 150n26. *See also* Allen rule; biogeographic "rules"
Bering Land Bridge. *See* Beringia
Bering Strait, 127, 134–136
Beringia, 4, 8, 95–97, 112, 115–117, 120–121
Biache, 50
Bickerton, Derek, 72–73, 160n15. *See also* language
bifaces, 25, 60, 112, 115. *See also* cleavers; hand axes; picks; points
Bilzingsleben, 43–44
biogeographic "rules," 81. *See also* Allen rule; Bergmann rule
bipedalism, 16–17, 24
birds, hunting of: Bel'kachinsk culture, 127; Beringia, 115, 117–118; Dorset culture, 133; Independence culture, 132; Late Stone Age (Norway), 123; modern humans (Upper Paleolithic), 70, 90, 92–94, 100, 111, 171–172n33; Neanderthals, 63–64; northern Russia, 108, Sumnagin culture, 112, 114
Birnirk phase, 136–137, 141–142
bison, steppe, 51, 58, 63, 65, 82, 95, 107, 110, 112, 114, 176n82
bitumen, 60. *See also* composite tools and weapons; hafting
blades and bladelets, stone, 88, 106, 126–127; end-blades, 129, 179n29; side-blades, 129
Blombos Cave, 75
Bluefish Caves, 115
boar, wild, 101, 107
boats: Komsa culture, 103; Late Stone Age (Norway), 123–124; Old Bering Sea/Okvik cultures, 136; Pre-Dorset culture, 132; settlement of Australia by, 89; Tasmanian, 165–166n74; Thule culture, 139; wooden oar for, 108. *See also umiak*
body hair, 10, 25, 38, 46, 54
bolas, 92
bombers, long-range, 141
bonobo. *See* chimpanzee
Border Cave, 84
boreal forest, 12, 107–109, 120, 128
Bose Basin, 147n57
bowhead whale. *See* whales and whaling
bows and arrows, 97, *108,* 108–109, 114, 136, 139; recurved bows, 3, 132–133; tanged arrowheads, 135
Boxgrove, 36–41, 46, 151n31
brachial index, *80,* 81, 109
brain, size of: apes, 10; australopithecines, 15; *Homo erectus*, 23–24, 27; *Homo habilis*, 18; *Homo heidelbergensis*, 36–37; modern human, 72; Neanderthal, 52–53
Broken Hill, 36
Broken Mammoth, 115–117, *116*
Bronze Age, 109, 125, 128, 169n1
Brooks, Alison, 73
Buret', 89, 94
burial of the dead: Beringia, 116; Ipiutak culture, 137–138; modern human (Upper Paleolithic), 84, *85,* 161n25; Neanderthal, 48, 66–67, 158nn66–67; northern Russia, 109; ritual associated with, 84, 109, 137–138, 164n55. *See also* cemeteries; "grave goods"
burins, 88, 101, 106, 112, 115, 117, 132

Cagny-Cimetière, 46
Cagny l'Epinette, 39
calendar sticks, 84
canine fossa, 32
cannibalism, 32
Cape Baranov, 136
Cape Krusenstern, 135
Cape Lisburne, 139, 141–142
Carbonell, Eudald, 32
caribou (*Rangifer tarandus*), 118, 129, 132, 135–136, 139–140. *See also* reindeer
Carpathian Mountains, 33
Caucasus Mountains, 23, 39, 43, 57, 63–65, 67, 77, 152n40

caves and rock shelters, 30, 36, 40, 51, 57, 62, 158n66
cemeteries, 109, 136–137
central-place foraging, 21, 40, 65, 152n40
Ceprano, 32
Chauvet Cave, 162n41
Chernoozer'e, 110
Childe, V. Gordon, 90, 167n92
chimpanzee, 10–13, 16–17, 20–21; pygmy chimp (bonobo), 19; toolmaking of, 10–11, *11*, 144n7
choppers, 19, 31, 44–45. *See also* Oldowan industry
Choris phase, 135
Chukchi Sea, 135, 137, 141
Chukotka, 114–115, 117, 127, 135
Clactonian industry, 153n54
Clacton-on-Sea, 42, 45, 152n46
classification: colors, 84; plants and animals, 71, 85–86
cleavers, 25–26, 41. *See also* Acheulean industry
clothing: foraging peoples of high latitudes, 167n87; *Homo heidelbergensis*, 42, 46; Inuit, 3, 90; lice and, 166n79; modern human (Upper Paleolithic), 70, 84, *89,* 89–90, 109, 115, 166n74; Neanderthal, 61–63; Viking, 4
Cold War, in the circumpolar zone, 140–142
Combe Grenal, 57
Combe-Sauniere I, 97
combs, 123
composite tools and weapons, 53, 60–61, *61,* 156n41. *See also* hafting
Coon, Carleton, 53–55, 59, 68
"core tools," 19, 23. *See also* Oldowan industry
Crimea, 77
Cro-Magnon, 81
crural index, 109. *See also* brachial index

Dansgaard-Oeschger events, 78, *79*
Danube Basin, 7, 28, 44, 94, 155n29
Darwin, Charles, 11, 47–48
Dederiyeh Cave, 59
deer, 32, 34, 58, 63, 101, 110, 112, 151n31, 173n50; *Dama nesti,* 23

Deetz, James, 88
Denali complex, 117, 119, 176nn82–83
Denbigh Flint complex, 127–132, 135, 137, 178n24
desert, 12, 87
Desna River, 58
Devon Island, 133
diatoms, 98
diet: adaptation to cold, 47; australopithecines, 17–18, 146n29; foraging peoples in higher latitudes, 27, 64, 158n58, 172n39; *Homo erectus,* 26–27; *Homo habilis,* 20–21; Inuit, 4–5; modern apes, 11; modern human (Upper Paleolithic), 70, 90, 92; Neanderthal, 8, 59, 62–64; Viking, 4
Diring Yuriakh, 155n33
Distant Early Warning (DEW) system, 142
Dmanisi, 23
DNA, 12, 48, 57, 68, 166n79
Dnestr Valley, 51, 157n49
dog, domesticated, 106, 108, 112; traction, 132, 139
dolina. *See* "Gran Dolina"
Don River, 76–77, 92, 94
Dorset culture, 1, 3, 5, 133–134, 136, 140, 181n63
Douglas, Mary, 85
driftwood, 104, 132
drills, 108
Dry Creek, 117
Dvuglazka Cave, 58
Dyuktai Cave, *111,* 112
Dyuktai culture, 112, 115, 117

East Siberian Sea, 114
Ehringsdorf, 50
Ekven, 136
El Castillo, 77
electron spin resonance (ESR) dating, 75, 161n22
elephant, 32, 34, 40
Eliseevichi, 106, 171n33
elk or red deer (*Cervus* sp.). *See* deer
Ellesmere Island, 1, 3
enamel: coating on teeth of, 11, 13–15, 17; hypoplasias, 100

entrance tunnels to houses, 117, *118*, 135, 139
Epi-Gravettian, 104–107, 109–110
Eskimo. *See* Inuit
Ethiopian plateau, 22
"executive brain," 72
external nose, 25

Finnmark county (Norway), 101
fire, use of controlled: *Homo erectus*, 26–27, 147–148n61; *Homo heidelbergensis*, 7, 42–43, 45–46, 152n46; modern human (Upper Paleolithic), 106; Neanderthal, 62; technology for making, 62, 108, 157n50, 167n87. *See also* Swartkrans Cave; Zhoukoudian
fired ceramics (pottery): Alaskan, 135–137; check-stamped, 135; Comb style, 124; cord-marked, *127;* earliest production of, 71, 91; Late Stone Age (Norway), 122, 124, 178n17; net-impressed, 125–126; Siberian Neolithic, 125–128, *127;* tempered with asbestos, 124
fish and fishing: Alaska, 118, 135–136; Beringia, 115, 117; fish hooks, 90, 112, 123; Independence culture, 132; Late Stone Age (Norway), 123–124; modern human (Upper Paleolithic), 70, 90, 92–93, 100, 106, 171n33; Neanderthal, 64; northern Russia, 108–109; Siberia, 111–112; Siberian Neolithic, 127; Thule culture, 140
floats: bark, 108; drag, 136
Florisbad, 72
food-sharing, 21, 25
foraging: *Homo erectus*, 7, 25–27; *Homo habilis*, 20–21; *Homo heidelbergensis*, 43; modern human (Upper Paleolithic), 90, 92–93, 100; Neanderthal, 65–66. *See also* central-place foraging; "home bases"
foraminifera, 34, 98
Fosna-Hensbacka complex, 101
fox, polar (or arctic) (*Alopex lagopus*), 5, 58, 78, 90, 92, 103, 106, 132, 139
Foxe Basin, 132
FOXP2 gene, 73. *See also* language

Franco-Cantabria, 57–58
Frere, John, 41
frontal lobe, 72
frostbite, 54, 95
fuel: animal fat, 92; bone, 92, 104–105; sea-mammal oil, 132–133, 139; wood, 94, 110, 115; *See also* fire, use of controlled; lamps, portable

gaffs, 123
Gagarino, 92
Gamble, Clive, 43
gazelle (*Gazella borbonica*), 23
Geissenklösterle, 83
gene flow, 68
genetic drift, 49, 53
Gesher Benot Ya'aqov, 36
giant cheetah (*Acinonyx pardensis*), 38
Giddings, J. Louis, 127, 130, 135, 179n29
Gjessing, Gutorm, 101–103
glacials, 34–35, 46–47
glaciers, 71, 96; Fennoscandian ice sheet, 101; retreat of, 116, 121, 128
goat, 63, 65
Goodall, Jane, 11
gorges, fish, 132
gorilla, 10–13
"Gran Dolina," 32. *See also* Atapuerca
"grave goods," 66, 109, 136, 158n67. *See also* burial of the dead
Gravettian industry, 91–94, 100–101, 103–104, 106, 134
Greenland, 1–4, 50, 78, 121, 130–134, 139, 141–142, 143n9. *See also* Independence culture
Greenland Ice-Core Project (GRIP), 78, 98
Greenland Ice Sheet Project 2 (GISP2), 78, 98
Grotta dei Moscerini, 64
Grotte des Enfants, 81
Grotte du Lazaret, 67
ground stone artifacts, 123; of slate, 122, 135–137
group size, 67–68
Gulf Stream current (North Atlantic Ocean), 101, 121
Guthrie, Dale, 114

hafting, 53, 60
hand axes, 23, 25–26, *26,* 41, 147n57, 150n20, 153n54; absence of, 30, 43, 45, 53, 59; function(s) of, 25–26. *See also* Acheulean industry
hare, 53, 90, 92, 111, 132
harpoons, barbed, 73, 99, 108, 111, 123; toggle-head or toggling, 3, 9, 129, 132–133, *133,* 135–136, 138
Hatt, Gudmund, 167n87
hearths. *See* fire, use of controlled
Heidelberg jaw, 36, 39, 49. *See also Homo heidelbergensis*
Heinrich layers and events, 78, *79*
hide working, 41–42, 46, 61–62, 89, 111
hippopotamus (*Hippopotamus amphibius*), 57
Holocene epoch, 130, 169n1
"home bases," 21, 40; hypothesis of, 21
hominid, 15, 18
hominoid, 13–15
Homo antecessor, 32, 149n13; *See also* Atapuerca
Homo erectus, 7, 32, 36, 44; anatomy, 23–25, *24,* 38, 72; foraging, 25; "protolanguage" and, 160n15; technology, 25–27; 160n15. *See also* Acheulean industry; hand axes; Nariokotome skeleton
Homo habilis, 18–24, 27, 39–40. *See also* Koobi Fora; Olduvai Gorge
Homo heidelbergensis, 7, 33, 45–47, 49, 52–53, 71–72, 149n13; anatomy, 36–38, *37;* diet, 38–40; foraging, 40, 65; technology, 41–43, 61. *See also* Boxgrove; Broken Hill; Heidelberg jaw; Petralona; Schoetensack, Otto
Homo sapiens. See modern humans
Homo neanderthalensis. See Neanderthals
honeybee dance, 160n12
horse (*Equus* sp.), 32, 34, 51, 63, 92, 95, 107, 114; *Equus stenonis,* 23
houses: Beringia, 116–118, *118;* Denbigh Flint complex, 129; Dorset culture, 133; Gravettian industry, 92; Gressbakken, 123; Ipiutak culture, 137; Karlebotn, 123; Komsa culture, 101, 103; Late Stone Age (Norway), 123; mammoth-bone (Epi-Gravettian), 104–105, *105,* 106, 110, 171n27; northern Russia, 109; Norton phase, 135; Old Bering Sea/Okvik cultures, 136; Old Whaling culture, 135; Thule culture, *139,* 139–140

Hoxne, 34, 39–42
Hudson Bay, 132, 140
Hultén, Eric, 115
hunting: chimpanzee, 11; Denali complex, 118; Denbigh Flint complex, 129; Dorset culture, 133; *Homo erectus,* 27; *Homo habilis,* 20; *Homo heidelbergensis,* 40; Independence culture, 132; Ipiutak culture, 137; Late Stone Age (Norway), 123–124; modern human (Upper Paleolithic), 92, 97, 100, 111; Neanderthal, 8, 63–65, 67; Norton phase, 135; Old Bering Sea/Okvik cultures, 136; Pre-Dorset culture, 132; Sumnagin culture, 112–114; Thule culture, 138–139
Huxley, Thomas Henry, 47
hyena, 27, 32
"hyperpolar," 53. *See also* anatomical adaptations to cold; Neanderthals

Ice Age. *See* Pleistocene epoch
"ice cellars," 92. *See also* storage pits
ice cores, 50–51, 78, *79,* 98, 130
ice-free corridor (northwest Canada), 116–117
Igloolik, 133
Independence culture, 130–133
Independence Fjord, 131
Indigirka River, 175n72
infraorbital foramina, 55
Ingalik, 172n39
intercontinental ballistic missile (ICBM), 142
interglacials, 33–35, 38, 51, 154n15, 155n33
interstadials, 78–79, 155n32, 171n23
Inuit (Eskimo), 103–104, 134–135, 138–140, 142, 181n63; adaptations to cold climate, 3–5, 54–55, 71, 92–93; Vikings and, 3–5. *See also* Thule culture
Ipiutak culture, 137–138, 178n17

Iron Age, 124–125, 128
Isaac, Glynn, 20–21, 88
Isernia La Pineta, 39, 45
Isturitz, 83
Iyatayet, 129, 135

Jebel Irhoud, 72
Journal of Human Evolution, 73

Kamchatka, 116–117
Kanzi (pygmy chimp), 19
Kara-Bom, 77
Katanda, 73
Khotylevo, 58
Khryashchi, 155n29
kilns, 91. *See also* fired ceramics
Kimeu, Kamoya, 24
King, William, 47
Kjelmoy, 124–125
Klein, Richard, 73
knuckle-walking, 10, 13
Knuth, Count Eigil, 130–132
Kobuk River, 129
Kokorevo, 110
Kola Peninsula, 101
Komsa culture, 101–103, 121, 123
Koobi Fora, 18, 27
Korolevo, 44, 155n29
Kostënki, 76–77, *77,* 81–83, 86, 92, *93,* 104, 166n79, 171n33
Kotzebue Sound, 135
Ksar 'Akil, 76
Kulichivka, 84, 89
Kunda culture, 107

Labrador, 1, 140
La Cotte de St. Brelade, 63, 157n56
Lagar Velho, 68
Lake Baikal, 110
Lake Mungo, 76
Lake Onega, 109
Lake Turkana, 22, 24
lamps, portable, 70, 92, 103, 132–133, 135, 139
language, 48, 66, 70–71, 72–73, 82–83; "proto-language," 160n15; relation to technology, 71, 87–88, 90, 165n71. *See also* anatomy of speech; modern humans

Laptev Sea, 114
Lartet, Edouard, 97
larynx. *See* vocal tract
Last Glacial Maximum, 91–97, 100, 103, 106, 168n99
Last Interglacial climatic optimum, 51, 57–58, 98, 155n29
Late Dorset. *See* Dorset culture
Lateglacial interstadial, 98, *99*
Late Stone Age (Norway), 121–125, 134, 178n17
Laugerie-Haute, 97
Leakey, Louis and Mary, 18
Lehringen, 60. *See also* wood technology
leisters, 108, 123, 132
Lena Basin, 112, 115, 125, 155n33, 175n72
lesser scimitar cat (*Homotherium latidens*), 38
Levallois technique. *See* prepared-core techniques
Linnaeus, 10
"Little Ice Age," 4, 130–131, 140
Liujiang, 76
loess, 34, 51, 91, 103, 115
long-distance movement (of materials and people), 68, 73, 86–87, 134, 164n64
"longhouses," 134, 168n97
Lower Paleolithic, 19, 23, 25, 40, 153n54, 155n33, 165n74. *See also* Acheulean industry; Clactonian industry; Oldowan industry
Lubbock, Sir John, 167n92
luminescence (TL) dating, 75, 155n33, 161n22

Maastricht-Belvédère, 154n15
Magdalenian industry, 97–101, 106–107
Malinowski, Bronislaw, 87
Mal'ta, 84, 94
mammoth, woolly (*Mammuthus primigenius*), 51, 58, 63, 92, 94–95, 104–107, 114, 157n56
Mamontovaya Kurya, 81
maritime economy, arctic, 9, 114, 121–123, 132, 135, 178n25
Markkleeberg, 46

masks: Dorset culture, 133, Ipiutak culture, *138*
mattocks, 92, 114
Mbuti (Pygmies), 157n50
McBrearty, Sally, 73
McBurney, Charles, 45
meat consumption, 20–21, 27, 38–40, 46, 62–63
Medieval Warm Period, 1, 4, 130–131, 138, 143n1
Mesolithic, 169n1
Mezhirich, 104–106, *105,* 171n33
Mezin, 90
Mezmaiskaya Cave, *65*
microblades, 110–112, 115, 117, 122, 127, 132
microwear analysis of artifacts, 26, 41–42, 60–62, 89, 92, 146n35
Middle Paleolithic, 40, 169n2. *See also* Mousterian industry
middle Upper Paleolithic, 91. *See* Gravettian industry; Solutrean industry
Miesenheim I, 39
Mikhailovskoe, 155n29
Miocene apes, 13–15, *14,* 88; *Dryopithecus,* 14; *Gigantopithecus,* 15; *Lufengpithecus,* 15; *Oreopithecus,* 15; *Pliopithecus,* 14; *Proconsul,* 13; *Sivapithecus,* 14
Miocene epoch, 13–15, 17, 28
Mochanov, Yuri, 112, 155n33
modern humans (*Homo sapiens*), 8, 26, 32, 36, 38–42, 47–48, 51–52, 59, 61, 66, 150n26; abandonment of areas during Last Glacial Maximum, 71, 94–95; anatomy, 71–73, 80–81, 95; dispersal out of Africa, *74,* 75–82, 85–86, 89, 95, 161n22, 162n34, 169n2; ecological niche, 90; extinction of Neanderthals and, 68–69, 81–82, 140, 159n79; origins, 71–75. *See also* anatomy of speech; language; Upper Paleolithic
molecular research: "biomolecular clock," 13; human origins, 11, 13, 15; modern humans, 68, 76; Neanderthals, 48–49, 57, 68; origins of speech, 73
molluscs, 98, 130
Molodova I, 157n49

moose or elk (*Alces alces*), 107, 110, 112, 118, 123–124, 127, 173n50
Mousterian industry, 53, 57–58, 60, 154n15, 155n29
Movius, Hallam, 43
"Movius Line," 43–44, *44,* 59
music and musical instruments, 66, 71, 73, 83, 163n47, 165n72
musk ox, 5, 103, 132

Nariokotome skeleton, *24,* 24–25, 38. *See also Homo erectus*
Neanderthals, 5, 8, 26, 28, 30, 33, 36, 45, 84, 153n2; abandonment of areas during the Lower Pleniglacial, 58, 155n32; anatomy, 32, 51–55, 159n79; burial of dead, 66–67, 158nn66–67; chewing complex, 55; diet, 38–40, 62–64, 156n36; evolution, 48–53, 71–72; extinction, 68–69, 91, 140; foraging, 64–66; geographic distribution and expansion, 57–59, 76, 81–82, 89, 155n33; technology, 41, 59–62. *See also* anatomical adaptations to cold; Atapuerca
Neander Valley, 47, 49, 57, 153n2
needles, sewing: absence prior to Upper Paleolithic, 42, 61; Beringia, 117; cases, 92, 107, 136; earliest specimens, 166n79; Late Stone Age (Norway), 123; modern human (Upper Paleolithic), 89, 92, 95, 98, 106, 111; northern Russia, 107. *See also* clothing
Nenana Valley, 117
Nenets, 125
Neo-Eskimo, 134. *See also* Birnirk phase; Old Bering Sea and Okvik cultures; Thule culture
Neolithic, 109, 125, 169n1; Siberian, 125–128, 135
nets, 64, 90; landing, 108; net-sinkers, 92, 135–136
networks, social or alliance, 87, 93–94, 105, 110, 134
"neural hypothesis," 73
Nganasans, 178n25
Nihewan Basin, 23
Nizhnee Veret'e, 107
Norse. *See* Vikings

North Pole, 1, 141
Northern Archaic tradition, 128
Norton phase, 135–137
Norton Sound, 127, 129, 135

Ob' River, 110
occipital bun, 52. *See also* Neanderthals
"oceanic effect," 6, 28–29, *29,* 100, 148n2
OIS (oxygen-isotope stage). *See* oxygen-isotope record
Okhotsk, Sea of, 127
Old Bering Sea and Okvik cultures (or phases), 135–138, *137*
Old Whaling culture, 135
Old World monkeys, 11, 13–15
Oldowan industry, 19, *20,* 23, 26, 30, 146n35
Olduvai Gorge, 18, 23, 27
Oleneostrovskii, 109
Omo, 72
Onion Portage, 129, 131, 135
orangutan, 10–13, 20
Orce Basin, 32
Origin of Races, The (Coon), 53
Origin of Species, The (Darwin), 47
ornaments, personal, 66, 73, 76, 84–85, *85,* 88, 116
Osgood, Cornelius, 172n39
Oswalt, Wendell, 143n5
oxygen-isotope record, 34, *35, 50,* 50–51, 78, *79,* 130; *See also* ice cores

Pääbo, Svante, 48
pair-bonding, 25, 67
Paleo-Eskimo, 1, 134, 139
paleomagnetism, 76; Brunhes/Matuyama boundary, 149n12
Paranthropus, 18, 21. *See also* australopithecines
Paviland, 81
Pearyland, 130–131
Pech de l'Azé, 57
Pechora Basin, 109
"pencil-like" (*karandashevid'nii*) cores, 112, 126
permafrost, 114
Petralona, 36
"phyletic gradualism," 48

physiological responses to cold, 5, 46, 53–55
picks, 25–26, 41. *See also* Acheulean industry
pigment, use of, 73
Pleistocene epoch, 169n1; Early, 21; Middle, 21, 34, *35, 50,* 50–51, 130, 155n33; Late, 50, 78, *79,* 96, 98, *99,* 130. *See also* Last Glacial Maximum; Pleniglacial
Pleniglacial: Lower, 51, 57–58, 68, 71, 76, 78, 91, 155n32; Middle, 51, 78, 81–82, 175n72; Upper, 71, 77–79, 90, 96, 115
Plesetsk, 142
Pliocene epoch, 15, 169n1
Point Hope, 137–138. *See also* Ipiutak culture
points: barbed bone, 126, 132, 135; bifacial, 117, 126–127; foliate, 95; Levallois, 60; shouldered, 91–92; slotted or grooved bone or antler, 106, 110–111, 114, 117; side-notched, 128; stemmed, 116; tanged, 101, 107
polar bear, 3, 114
polar fox. *See* fox, polar
pollen cores and samples, 51, 78, 98, 110, 114, 130, *131*
polyhedrons, 31
Popovo, 109
porpoise, 123
pottery. *See* fired ceramics
Pre-Dorset culture, 132–133
prepared-core techniques, 53, 59–61, *61. See also* Mousterian industry
Prezletice, 31
productivity, plant and animal, 6, 22, 146n42; Last Glacial Maximum, 91, 103; measurement of, 144n13; oceanic effect and, 28; scavenging and, 40; Siberia, 110, 112; tropical zone, 12; tundra zone, 113, 120, 144n13
Proto-Denbigh culture, 131
proxemics, 163n50
"punctuated equilibria," 48
Punuk phase, 136–137
Pyrenees, 43

Qafzeh, 76
"Quest for Fire," 62, 157n50. *See also* fire, use of controlled; Neanderthals

radar stations in circumpolar zone, 141–142
Radcliffe-Brown, A. R., 87
"radiator nose theory," 55. *See also* Coon, Carleton; Neanderthals
radiocarbon dating, 75, 82, 94–95, 103, 110, 114, 162n41; calibration of, 159n1
red deer. *See* deer
red ochre, 84, 164n55
reindeer (*Rangifer tarandus*), 57–58, 63, 92, 94, 97, 101, 106–107, 110, 112, 114, 123, 127, 152n40; herding, 120, 125, 178n25
retromolar space, 49
Rhine Valley, 39
Rightmire, G. Philip, 149n13
rock shelters. *See* caves and rock shelters
roe deer. *See* deer
runestone, 1, *2*

Saami (Lapps), 103, 124–125
saiga, 63
St. Lawrence Island, 135–136
Sautuola, Don Marcelino Sanz de, 82
scanning electron microscope (SEM), 19, 39
scavenging, 20, 25, 27, 38, 40, 63–64, 157n55
Schoetensack, Otto, 36
Schöningen, 39–40, *42,* 43–44
Scladina Cave, 62–63
scrapers, stone, 19, 60, 101, 112
Scytho-Siberian style, 138. *See also* Ipiutak culture
sea surface temperatures (SST), 98
seal, 4, 114, 123–124, 129, 132–133, 135–137, 139, 179n29
seasonality, 6, 22
sedentism, 123, 125
SEM. *See* scanning electron microscope
Severskii Donets River, 155n29
sexual dimorphism, 10, 18–19, 25
sexual division of labor, 67

shamanism, 133, 138
Shanidar Cave, 59, 158n67
sheep, 63, 65, 117–118
shellfish, 64, 100
shells, 86, 105. *See also* long-distance movement; ornaments, personal
shelters, artificial, 42, 62, 70, 89–90, 110, 157n49
Siberia, 4, 6, 8, 134; Cold War in, 141; modern human (Upper Paleolithic) settlement of, 70–71, 76–78, 81, 86, 89, 91, 94, 96, 100, 110–112, 115, 168n99; Neanderthal settlement of, 47, 58, 62
Siberian Neolithic. *See* Neolithic, Siberian
Sima de los Huesos ("Pit of the Bones"), 49. *See also* Atapuerca; Neanderthals
Singa, 72
Skhul, 76
skis, 108, *123*
skraeling, 1, 143n3
slat armor, 136
sleds or sledges, 3, 108, 114, 133, 136, 139
small mammals, 63–64, 70, 90
snares. *See* traps and snares
snow goggles, 136
snowhouses, domed, 133–134
snow knives, 133
Society of Antiquaries, 41
Solutrean industry, 94–95, 97–98, 100
Soviet Union, 141–142
spear-throwers, 97, 106
stable isotope analysis of bone and teeth, 38, 62–64, 90, 92, 106, 146n29
Steinheim, 50
Steward, Julian H., 87
stone tool marks on bones, 19–20, 32, 39, 41, 45, 63, *64,* 151n31, 157n56
storage pits, 92–93, *93, 94,* 104–106, 110
Strait of Hormuz, 18
Studenoe, 110
Subarctic, 70, 125, 140; defined in terms of climate, 170n20
Sub-Atlantic period, 130–131
Sumnagin culture, 112–114, 118, 126
Sungir', 84, *85*
suprainiac fossa, 49
supraorbital torus, 52

Swan Point, 115
Swanscombe, 43, 50
Swartkrans Cave, 27
Syalakh culture, 125–127
symbolism, 66–67, 71, 73, 75, 82–88
syntax, 73, 88, 160n15, 165n74. *See also* language

Tagenar VI, 114
taiga. *See* boreal forest
Taimyr Peninsula, 114, 127
Tanana River, 115–116, 176n82
Tasmanians, 157n50, 165n74
technology: bone, antler, and ivory, 61, 84, 88, 92, 106, 111; *Homo erectus,* 25; *Homo heidelbergensis,* 41–43; innovative, 59–60, 70–71, 87–92, 97–98, 108–109; Inuit, 3; mechanical, 71, 88, 90, 97, 103, 107, 165n73, 167n87; modern human (Upper Paleolithic), 8, 41, 70, 75, 81, 87–91; Neanderthal, 59–62, 64; progressive improvement of, 109, 115; Saami, *124;* Viking, 3
temperature tolerance: *Homo erectus,* 22; *Homo heidelbergensis,* 33–34; modern apes, 12; modern human (Upper Paleolithic), 70–71, 78, 95; Neanderthals, 55–59; Yakuts, 169n108; Yukaghir, 95
Terra Amata, 42
Teshik Tash, 59
textiles, 71
throwing board, dart, 90, 136
Thule Air Base, *141,* 141–142. *See also* Cold War
Thule culture, 141–142; dispersal across North American Arctic, 138–140; origins, 134–138. *See also* Inuit
toggle-head harpoon. *See under* harpoons, barbed
Tolbaga, 166n79
tools. *See* technology
Torralba/Ambrona, 40
Toth, Nicholas, 19
traps and snares, 64, 70, 90, 92, 106, 108, 139
tree-ring data, 130
Treugol'naya Cave, 39, 152n40
trip alarm for winter ice hunting, 139, 181n61
tundra, 5, 12, 87, 109, 114–115, 117, 127; productivity of, 91, 113, 120, 144n13
Tunnit people, 181n63. *See also* Dorset culture
Turner, Alan, 38, 40, 46
Tyson, Edward, 10

Uelen, 136
Uivvaq, *139,* 141–142
ulu, 92
umiak, 3, 136, 138, 140
U.S. Air Force, 141
Upernavik, 1, *2*
Upper Angara River, 111
Upper Paleolithic, 88–89, 134, 167n92, 169n2. *See also* burins; Gravettian industry; Magdalenian industry; modern humans; Solutrean industry
Ushki, 116–117, *118*
Ust'-Belaya, 112

Varanger Fjord, 101, 123–125
"Venus" figurines, 87, 91, 94
Verkholenskaya Gora, 111
Vértesszöllös, 40, 43–44
Vézère River, 97
Vikings, 1–5, *2,* 143n9
Vindija Cave, 62
Vis I, 107–108, *108,* 112
vocal tract, 66, 72–73, *73,* 76. *See also* anatomy of speech

Walakpa, 129
walrus, 3, 114, 123, 129, 132, 134–136, 140, 179n29
warfare, 3, 136, 140; twentieth-century, 141–142
wedge-shaped microblade cores, 110–112, *111,* 117. *See also* microblades
weirs, fish, 64, *124,* 132
whales and whaling, 3, 123–125, 136–138; bowhead, 138, 140; whaling crews, 138. *See also umiak*
White, Leslie, 165n74
White Sea, 124–125
Willendorf, 77

"winged objects," 136. *See also* Old Bering Sea and Okvik cultures
wolf, 92
wood technology, 41–42, *42,* 46, 60–61, 107–109, 152n46
Wrangel Island, 135, 178n25

Xiaochangliang, 23

Yakuts, 169n108
Yana River, 169n108; RHS site, 81, 175n72

Yenisei River, 58, 110–111
Ymyyakhtakh culture, 128
Younger Dryas event, 98, *99,* 101, 107, 117–118
Yudinovo, 104–105
Yukaghir, 71, 95, 178n25
Yukon River, 172n39

Zaraisk, 90, 92, *94*
Zhokhov Island, 114, 121
Zhoukoudian, 148n61
Zveinieki, 107

About the Author

John F. Hoffecker is a Fellow of the Institute of Arctic and Alpine Research at the University of Colorado in Boulder. He received an M.A. from the University of Alaska in 1979 and a Ph.D. from the University of Chicago in 1986. Much of his research and writing has been focused on the ecology of past peoples in cold environments. He has conducted field investigations of early prehistoric sites in central Alaska, Neanderthal caves in the Northern Caucasus, early Upper Paleolithic sites on the central East European Plain, and late prehistoric Inuit occupation on the northwest coast of Alaska. He is the author of *Desolate Landscapes: Ice-Age Settlement in Eastern Europe* (Rutgers University Press, 2002).